P9-DJL-982

FEMINISM CONFRONTS TECHNOLOGY

In memory
of my father
Szloma Wajcman
1905–1978

FEMINISM CONFRONTS TECHNOLOGY

JUDY WAJCMAN

The Pennsylvania State University Press
University Park, Pennsylvania

First published 1991 in the United States by
The Pennsylvania State University Press,
Suite C, 820 North University Drive,
University Park, PA 16802

ISBN 0–271–00801–6 (cloth)
ISBN 0–271–00802–4 (paper)

Library of Congress Cataloging in Publication Data
Wajcman, Judy.
Feminism confronts technology/Judy Wajcman.
p. cm.
Includes bibliographical references (p.) and index.
ISBN 0–271–00801–6 (cloth): $25.00. —
ISBN 0–271–00802–4 (paper): $11.95
1. Technology—Social aspects. 2. Feminist criticism.
I. Title.
HM221.W35 1991
306.4′6—dc20 91–18539
CIP

It is the policy of The Pennsylvania State University Press to use acid-free paper for the first printing of all clothbound books. Publications on uncoated stock satisfy minimum requirements of American National Standard for Information Sciences—Permanence of Paper for Printing Library Materials, ANSI 239, 48-1984

Printed in Great Britain

Contents

Acknowledgements vi

Preface viii

1 Feminist Critiques of Science and Technology 1

2 The Technology of Production: Making a Job of Gender 27

3 Reproductive Technology: Delivered into Men's Hands 54

4 Domestic Technology: Labour-saving or Enslaving? 81

5 The Built Environment: Women's Place, Gendered Space 110

6 Technology as Masculine Culture 137

Conclusion 162

Bibliography 168

Index 181

Acknowledgements

In acknowledging the help I received in producing this book, I must begin with Jenny Earle. She not only read and expertly edited every chapter, but discussed with me all the ideas, intellectual and political. I am indebted to her for constant support and encouragement throughout the entire process.

I started out intending to write an article – it was Donald MacKenzie who suggested I might as well write a book. I have often cursed him since! Now it is finished I can thank him for his suggestion and for initially fostering my interest in this area. He and Tony Giddens have seen the book through from beginning to end, promptly reading its many drafts and exhorting me to continue through despondency and ill-humour. I am grateful to them both and relieved that our friendships have survived. Cynthia Cockburn also read many drafts and her belief in the project was likewise very important to me.

The research and writing for the most part took place while I was a visitor at the Department of Applied Economics at Cambridge University. As in the past, this was an excellent workplace for me. I would like to thank Bob Blackburn, who arranged my visit, Ken Prandy, Jeremy Edwards, Lucia Hanmer and Kath Wilson all of whom made me welcome and were a continuing source of encouragement.

While in Cambridge, I was most fortunate to live in a household with Ros Morpeth and Lol Sullivan. They were both extremely generous to me, from patiently engaging about the ideas in the book to fixing my bicycle! In particular, I am grateful to Ros for her warm hospitality and unswerving friendship. I am indebted to many other friends in Cambridge, London and Edinburgh and I would particularly like to mention Karen Greenwood, Lynn Jamieson, Archie Onslow, Mary Ryan and Michelle Stanworth. For their helpful comments on particular chapters and for inspiring me with their own work, I must thank Ruth Schwartz Cowan, Sandra Harding, Maureen McNeil and others associated with the UK Gender and Information Technology Network. The last stages of the manuscript were completed when I returned to Australia and I would like to thank Pauline Garde, Richard Gillespie, Carol Johnson, Belinda Probert, Stuart Rosewarne and especially Michael Bittman for helping me with the final draft.

I had the benefit of a small research grant from the University of New South Wales and the Australian Research Council. I am grateful to Alex Heron, Karen Hughes, Jocelyn Pixley, Alison Tilson and especially Mandy Wharton for their exemplary, if all too brief, research assistance.

This is by no means an exhaustive list of contributors to my thinking and writing on gender and technology. A project with as broad a sweep as this, extending over an intensive two-year period, has of course involved me in numerous conversations particularly with other feminists working in the area. Feeling part of a collective feminist endeavour has sustained me and I hope the end product usefully contributes to this ongoing work.

Preface

Over the last two decades feminists have identified men's monopoly of technology as an important source of their power; women's lack of technological skills as an important element in our dependence on men. From Women in Manual Trades, set up in the early 1970s to train women in traditionally male skills, to the Women and Computing courses of the 1980s, feminist groups and campaigns have attempted to break men's grip on technical expertise and to win greater autonomy and technical competence for women. In the same period, women's efforts to control their own fertility have extended from abortion and contraception to mobilizing around the new reproductive technologies. With dramatic advances in biotechnology and the prospect of genetic engineering, women's bodies have in some respects become increasingly vulnerable to exploitation.

These and other political struggles around technology, and the difficulties they continue to confront, have opened up an exciting new field in feminist scholarship. To date however, most contributions to the debate on gender and technology have been of a somewhat specialist character, focused on a particular type of technology. This book represents an attempt at a more coherent approach, bringing together under one theoretical framework a number of different sites of technology. It is my intention both to explicate and to extend the newly emerging feminist analysis.

Turning to social science debates about technology we find a preoccupation with the impact of technological change on society. Many commentators, for example, claim we are in the midst of a microelectronic revolution, which will cause a radically new form of society to emerge. Regardless of their theoretical or political perspectives, women rarely enter their field of vision. Feminists have worked to put women and gender relations back into this frame, highlighting the differential effects of technological change on women and men. Although still largely concerned with 'effects', feminists also point beyond the relations of paid production to a recognition that technology impinges on every aspect of our public and private lives. While I will be engaging with these issues, I also intend to take the analysis into less charted waters.

The technological determinism implicit in much of both the sociological and feminist literature on the impact of technology has recently been subjected to criticism. The new sociology of technology has turned the focus around to examine the social factors that shape technological changes. Rather than only looking at the effects of technology on society, it also looks at the effects of society on technology.

The Social Shaping of Technology (1985), which I co-edited with Donald MacKenzie, was part of this project. As an edited collection, that book was to some extent deficient in its treatment of gender issues, reflecting the state of knowledge at that time. This book is motivated by a desire to redress the balance, exploring in more depth women's relationship to and experience of technology. Rather than providing a comprehensive review of the now burgeoning literature in this area, I have selected research which can best exemplify the centrality of gender relations to the social shaping approach.

I have not attempted to encompass here all forms of technology. I have not, for example, dealt with the technologies of surveillance and political control, nor with energy technology. Various aspects of information and communication technologies have also been excluded. I have chosen to concentrate on advanced industrial societies, and the book has few references to the major issues concerning technology in the Third World. There is now an extensive literature on how technology transfer to the Third World has a powerful tendency to reinforce male dominance.[1] In the end, the sheer scope of the topic prohibited its inclusion.

The book begins with an overview of feminist theories of science and technology. In this first chapter, I argue that the feminist critique of science cannot simply be translated into a feminist perspective on technology. Although useful parallels can be drawn, technology needs to be understood as more than applied science. The following chapters have a less abstract focus and are organized around substantive areas of technology. Each chapter begins by looking at the impact of technological change on sexual divisions and goes on to develop the argument that technology itself is gendered.[2]

Chapter 2 assesses the impact of production technologies on sexual divisions in the sphere of paid work. It then looks at the extent to which these divisions, and gender relations in the workplace, themselves profoundly affect the direction and pace of technological change.

Perhaps it is the new technologies of human biological reproduction that have been most vigorously contested, both intellectually and politically, by feminists in recent years. Chapter 3 explores the arguments, placing them in the wider context of the growing supremacy of technology in Western medicine.

There is now a substantial body of feminist writing on domestic technologies and their bearing on housework. Chapter 4 examines this research in conjunction with more mainstream (malestream) sociological theories regarding the impact of technologies on the 'post-industrial' home.

Chapter 5 deals with the built environment. The first section considers the design of houses and their urban location. I argue that sexual divisions are literally built into houses and indeed into the whole structure of the urban system. The last section scrutinises transport technology and demonstrates how women in particular have been disadvantaged by the design of cities around the automobile.

Picking up on issues from the previous four chapters, chapter 6 presents an analysis of technology as a masculine culture. I argue that the close affinity between technology and the dominant ideology of masculinity itself shapes the production and use of particular technologies. The correspondingly tenuous nature of women's relationship to this technical culture is the subject of the second part of the chapter.

In the conclusion, I hope to convince the reader that a recognition of the profoundly gendered character of technology need not lead to political pessimism or total rejection of existing technologies. The argument that women's relationship to technology is a contradictory one, combined with the realization that technology is itself a social construct, opens up fresh possibilities for feminist scholarship and action.

NOTES

1 For an introduction to this literature, see McNeil's (1987, pp. 227–9) bibliography on 'Development, The "Third World" and Technology'. See also Ahmed (1985).

2 Throughout this book I use the term 'sex' and 'gender' interchangeably. This is symptomatic of the blurred boundaries that mark the distinction between what is construed as 'natural' and what is construed as 'social'.

1
Feminist Critiques of Science and Technology

Writing in 1844 about relations between men and women, Marx said that '[i]t is possible to judge from this relationship the entire level of development of mankind' (1975, p. 347). More commonly it is the level of scientific and technological development that is taken as the index of a society's advancement. Our icons of progress are drawn from science, technology and medicine; we revere that which is defined as 'rational' as distinct from that which is judged 'emotional'. As we approach the twenty-first century however we are no longer sure whether science and technology are the solution to world problems, such as environmental degradation, unemployment and war, or the cause of them. It is not surprising therefore that the relationship between science and society is currently being subjected to profound and urgent questioning.

The development of a feminist perspective on the history and philosophy of science is a relatively recent endeavour. Although this field is still quite small and by no means coherent, it has attracted more theoretical debate than the related subject of gender and technology. It will become apparent in what follows, however, that feminists pursued similar lines of argument when they turned their attention from science to technology. I will therefore start by examining some approaches to the issue of gender and science, before moving on to look at technology.

The Sexual Politics of Science

The interest in gender and science arose out of the contemporary women's movement and a general concern for women's position in the professions. Practising feminist scientists have questioned the historical and sociological relationships between gender and science at least since the early 1970s. The publication of biographical studies of great women scientists served as a useful corrective to mainstream histories of science in demonstrating that women have in fact made important contributions to scientific endeavour. The biographies of Rosalind Franklin and Barbara McClintock, by Anne Sayre (1975) and Evelyn Fox Keller (1983) respectively, are probably the best known examples. Recovering the history of women's achievements has now become an integral part of feminist scholarship in a wide range of disciplines. However, as the extent and intransigent quality of women's exclusion from science became more apparent, the approach gradually shifted from looking at exceptional women to examining the general patterns of women's participation.

There is now considerable evidence of the ways in which women have achieved only limited access to scientific institutions, and of the current status of women within the scientific profession. Many studies have identified the structural barriers to women's participation, looking at sex discrimination in employment and the kind of socialization and education that girls receive which have channelled them away from studying mathematics and science. Explaining the under-representation of women in science education, laboratories and scientific publications, this research correctly criticises the construction and character of feminine identity and behaviour encouraged by our culture.

However these authors mainly pose the solution in terms of getting more women to enter science - seeing the issue as one of access to education and employment. Rather than questioning science itself, such studies assume that science is a noble profession and a worthy pursuit and that if girls were given the right opportunities and encouragement they would gladly become scientists in proportion to their numbers in the population. It follows that remedying the current deficiency is seen as a problem which a combination of different socialization processes and equal opportunity policies would overcome.

This approach, as Sandra Harding (1986) and others have pointed out, locates the problem in women (their socialization, their aspirations and values) and does not ask the broader questions of whether and in what way science and its institutions could be reshaped to accommodate women. The equal opportunity recommendations, moreover, ask women to exchange major aspects of their gender identity for a masculine version without prescribing a similar 'degendering' process for men. For example, the current career structure for a professional scientist dictates long unbroken periods of intensive study and research which simply do not allow for childcare and domestic responsibilities. In order to succeed women would have to model themselves on men who have traditionally avoided such commitments. The equal opportunities strategy has had limited success precisely because it fails to challenge the division of labour by gender in the wider society. The cultural stereotype of science as inextricably linked with masculinity is also crucial in explaining the small number of women in science. If science is seen as an activity appropriate for men, then it is hardly surprising that girls usually do not want to develop the skills and behaviours considered necessary for success in science.

When feminists first turned their attention to science itself, the problem was conceived as one of the uses and abuses to which science has been put by men. Feminists have highlighted the way in which biology has been used to make a powerful case for biologically determined sex roles. Biology has been central to the promotion of a view of women's nature as different and inferior, making her naturally incapable of carrying out scientific work. For example, sex differences in visual-spatial skills are said to explain why there are so many more male scientists. In confronting biological determinists, many feminists inquired as to how and why the study of sex differences had become a priority of scientific investigation. They set out to demonstrate that biological inquiry, and indeed Western science as a whole, were consistently shaped by masculine biases. This bias is evident, they argued, not only in the definition of what counts as a scientific problem but also in the interpretations of research. It followed that science could not be genuinely objective until the masculine bias was eliminated. As we shall see below, this approach leaves unchallenged the existing methodological norms of scientific inquiry and identifies only bad science and not science-as-usual as the problem.

The radical political movements of the late 1960s and early 1970s also began with the question of the use and abuse of science. In their campaigns against an abused, militarized, and polluting science they argued that science was directed towards profit and warfare. Initially science itself was seen as neutral or value-free and useful as long as it was in the hands of those working for a just society. Gradually, however, the radical science movement developed a Marxist analysis of the class character of science and its links with capitalist methods of production. A revived political economy of science began to argue that the growth and nature of modern science was related to the needs of capitalist society. Increasingly tied to the state and industry, science had become directed towards domination. The ideology of science as neutral was seen as having a specific historical development. One of the most characteristic formulations of this position, associated with the radical science movement, was that 'science is social relations'.

The point was that the distinction between science and ideology could not be sustained because the dominant social relations of society at large are constitutive of science.

During this same period a radical shift took place in the history, philosophy and sociology of science, which added weight to the view that science could no longer be understood simply as the discovery of reality. Thomas Kuhn's *The Structure of Scientific Revolutions* (1970) marked the beginning of what was to become a major new field of study known as the sociology of scientific knowledge.[1] Its central premise is that scientific knowledge, like all other forms of knowledge, is affected at the most profound level by the society in which it is conducted.

Much research has examined the circumstances in which scientists actually produce scientific knowledge and has demonstrated how social interests shape this knowledge. Studies provide many instances of scientific theories drawing models and images from the wider society. It has also been demonstrated that social and political considerations enter into scientists' evaluations of the truth or falsity of different theories. Even what is considered as 'fact', established by experiment and observation, is social. Different groups of scientists in different circumstances have produced radically different 'facts'. Numerous historical and contemporary studies of science, and the social processes through which inquiry proceeds, highlight the social aspects of scientific knowledge.

Despite the advances that were made through the critique of science in the 1970s, gender-conscious accounts were rare. The social studies of natural science systematically avoided examining the relationship between gender and science in either its historical or sociological dimensions. Similarly, the radical science movement focused almost exclusively on the capitalist nature of science ignoring the relationship of science to patriarchy. In short, gender did not figure as an analytical tool in either of these accounts of science.

It is only during the last decade with writers such as Carolyn Merchant (1980), Elizabeth Fee (1981), Evelyn Fox Keller (1985), Brian Easlea (1981), Nancy Hartsock (1983), Hilary Rose (1983) and Ludmilla Jordanova (1980) that Western science has been labelled as inherently patriarchal.[2] As Sandra Harding (1986) expresses it, feminist criticisms of science had evolved from asking the 'woman question' in science to asking the more radical 'science question' in feminism. Rather than asking how women can be more equitably treated within and by science, they ask 'how a science apparently so deeply involved in distinctively masculine projects can possibly be used for emancipatory ends' (p. 29). It is therefore time to consider the main feminist critiques of science itself.

Scientific Knowledge as Patriarchal Knowledge

The concern with a gender analysis of scientific knowledge can be traced back to the women's health movement that developed in Britain and America during the 1970s. Regaining knowledge and control over women's bodies - their sexuality and fertility - was seen as crucial to women's liberation. Campaigns for improved birth control and abortion rights were central to the early period of second-wave feminism. There was a growing disenchantment with male medical theories and practices. The growth and consolidation of male expertise at the expense of both women's health and women's healing skills was the theme of an American study, *Witches, Midwives and Nurses: A History of Women Healers* (Ehrenreich and English, 1976). This documented how the growth and professionalization of male-dominated medicine had led to the marginalization of female health workers. At the same time, critiques of psychiatry and the treatment of women's depression as pathological were being expounded. Asking why the incidence of mental illness should be higher among women than men, feminists exposed the sexist bias in medical definitions of mental health and illness. Implicit in these analyses was a conviction that women could develop new kinds of knowledge and skills, drawing on their own experience and needs. The insights of the radical science movement contributed to the view of medical science as a repository of patriarchal values.

If medical scientific knowledge is patriarchal, then what about the rest of science? As Maureen McNeil (1987) points out, it was a short step to the emergence of a new feminist politics about scientific knowledge in general. Some feminists re-examined the Scientific Revolution of the sixteenth and seventeenth centuries, arguing that the science which emerged was fundamentally based on the masculine projects of reason and objectivity. They characterized the conceptual dichotomizing central to scientific thought and to Western philosophy in general, as distinctly masculine. Culture vs. nature, mind vs. body, reason vs. emotion, objectivity vs. subjectivity, the public realm vs. the private realm - in each dichotomy the former must dominate the latter and the latter in each case seems to be systematically associated with the feminine. The general issue of whether conceptual dichotomizing is itself distinctly masculine or part of the Western philosophical tradition is beyond the scope of this book.[3] My concern is with the way dualistic gender metaphors such as those used above reveal the underlying social meanings in purportedly value-neutral scientific thought.

There has been a growing awareness of the use of female metaphors for nature and natural metaphors for women. An examination of the texts of science highlights the correspondence between the way men treated women in particular historical periods and the way they used nature. Some feminist historians have focused on the rape and torture metaphors in the writings of Sir Francis Bacon and the other fathers of modern science. Merchant (1980) argues that during the fifteenth to seventeenth centuries in Europe both nature and scientific inquiry were conceptualized in ways modelled on men's most violent and misogynous relationships to women and this modelling has contributed to the distinctive gender symbolism of the subsequent scientific world view.

Eighteenth and nineteenth century biomedical science in France and Britain deployed similar gender symbolism to conceptualize nature: '. . . science and medicine as activities were associated with sexual metaphors which were clearly expressed in designating nature as a woman to be unveiled, unclothed and penetrated by masculine science' (Jordanova, 1980, p. 45). Anatomically, males were depicted as representing active agents and females as passive objects of male agency. From her study Jordanova concludes that biomedical science intensified the cultural association of nature with passive, objectified femininity and of culture with active, objectifying masculinity. This strikingly gendered imagery of nature and of scientific inquiry is not just an historical relic, as these same dichotomies and metaphors can be found in contemporary writing on science. As Harding asks, is it any wonder that women are not an enthusiastic audience for these interpretations?

Rather than pointing to the negative consequences of women's identification with the natural realm, some feminists celebrate the identification of woman and nature. This finds political expression in the eco-feminism of the eighties which suggests that women must and will liberate the earth because they are more in tune with nature. For them, women's involvement in the ecology and peace movements was evidence of this special bond. As Susan Griffin expressed it: 'those of us who are born female are often less severely alienated from nature than are most men' (1983, p. 1). Women's biological capacity for motherhood was seen as connected to an innate selflessness born of their responsibility for ensuring the continuity of life. Nurturing and caring instincts are essential to the fulfilment of this responsibility. Conversely, men's inability to give birth has made them disrespectful of human and natural life, resulting in wars and ecological disasters. From this perspective, a new feminist science would embrace feminine intuition and subjectivity and end the ruthless exploitation of natural resources. Rejecting patriarchal science, this vision celebrates female values as virtues and endorses the close relationship between women's bodies, women's culture and the natural order.

While eco-feminism sees women's values as having a biological basis, another approach to the question of women and science has been informed by psychoanalysis. The object–relations school of thought has been particularly influential in the feminist conceptualizations of science. This theory describes the mechanisms through which adult women and men come to model themselves and their relation to the world in different ways. To acquire his masculine identity the boy must both reject and deny his former dependencies, attachment and identification with the mother. The resulting conflicts in men over masculinity create a psychology of male dominance.

Using this theory Keller argues that girls and boys have different cognitive skills. As the male distinguishes himself from the mother, he also learns to differentiate sharply between subject and object, between himself and others. According to Keller, as scientists are men this male mind set, obsessed with detachment and mastery, has been written into the norms and methods of modern science. A radically different scientific method is described by Keller (1983) in her influential biography of Barbara McClintock. A Nobel prize-winning geneticist, McClintock is described as a scientist who merged subject and object in her 'feeling for the organism' and whose work was imbued with a holistic understanding of, and reverence for, nature. According to Keller, this woman's work provides us with 'a glimpse of what a gender-free science might look like' by combining masculine and feminine characteristics. Rather than celebrating a woman-centred science as do the eco-feminists, this project insists on the possibility of a gender-neutral science produced by androgynous individuals.[4]

While emphatically rejecting the possibility of a neutral objective science, other feminist writers have shared a concern with the exclusion of woman-centred values from science. However, they attribute such values not to the individual psyche but to a socially and historically constructed gender division of labour. They trace the way in which, as the spheres of public and private life became increasingly separated during the course of the eighteenth century, women became confined to the private sphere of hearth and home. Skills such as reasoning and objectivity became associated with public life, and feeling and subjectivity with private life. These dichotomies have become historically associated with the development of distinctive feminine and masculine worldviews.

In a well-known article, Rose (1983) locates herself within the radical science tradition and endorses the Marxist characterization of bourgeois science as a form of alienated and abstract knowledge. It is the division of mental and manual labour, integral to capitalist production, which gives rise to this form of knowledge. Rose takes issue with this tradition however for its failure to question the impact of the gender division of labour on science. The focus of the radical science critique on the relations of production to the exclusion of reproduction negates women's experience, which in turn impoverishes science. Science has been denied the input of women's experience of the caring, emotionally demanding labour which has been assigned exclusively to women. According to Rose, a feminist science would need to encompass this emotional domain and thereby fuse subjective and objective ways of knowing the world. It would thus be a more complete, truer knowledge because it is based on women's 'shared experience of oppression'. Rose concludes that the reunification of 'hand, brain *and* heart' would foster a new form of science, enabling humanity to live in harmony with nature.

A Science Based on Women's Values?

These debates about science mirror the more general preoccupations that have engaged feminists over the last two decades. Much early second-wave feminism was of a liberal cast, demanding access for women within existing power structures, such as science. In principle, equality could be achieved by breaking down gender stereotypes: for instance by giving girls better training and more varied role models, and by introducing equal opportunity programmes and anti-discrimination legislation. Such feminist writing focused on gender stereotypes and customary expectations, and denied the existence of any fundamental sex differences between women and men. This first approach, liberal feminism, was based on an empiricist view of science as (gender) neutral. Sexism and androcentrism were therefore conceived of as social biases correctable by stricter adherence to the existing methodological norms of scientific inquiry. I would argue that the limitations of this approach have been made apparent by the sociology of scientific knowledge and the profound critique of empiricism that has occurred in the last few decades.

By the late 1970s however a new form of radical feminism, or cultural feminism as it is known in North America, had emerged which exalted femininity for its own sake. These writers emphasize gender difference and celebrate what they see as specifically feminine, such as women's greater humanism, pacifism, nurturance and spiritual development. Some of these authors abandoned the idea that what was 'specifically feminine' was socially produced and notions of ineradicable difference have flourished.

This return to an emphasis on natural or psychological gender difference is a common thread in many of the feminist views of science. They promote women's values as an essential aspect of human experience and seek a new vision of science that would incorporate these values. At this juncture therefore, I think it appropriate to point to some fundamental problems with the general assertion of a science based on women's values.

Essentialism, or the assertion of fixed, unified and opposed female and male natures has been subjected to a variety of thorough critiques.[5] The first thing that must be said is that the values being ascribed to women originate in the historical subordination of women. The belief in the unchanging nature of women, and their association with procreation, nurturance, warmth and creativity, lies at the very heart of traditional and oppressive conceptions of womanhood. 'Women value nurturance, warmth and security, or at least we believe we ought to, precisely because of, not in spite of, the meanings, culture and social relations of a world where men are more powerful than women' (Segal, 1987, p. 34). It is important to see how women came to value nurturance and how nurturance, associated with motherhood, came to be culturally defined as feminine within male-dominated culture. Rather than asserting some inner essence of womanhood as an ahistorical category, we need to recognize the ways in which both 'masculinity' and 'femininity' are socially constructed and are in fact constantly under reconstruction.

Secondly, the idea of 'nature' is itself culturally constructed. Conceptions of the 'natural' have changed radically throughout human history. As anthropologists like Marilyn Strathern and others have pointed out, 'no single meaning can in fact be given to nature or culture in Western thought; there is no consistent dichotomy, only a matrix of contrasts' (Strathern, 1980, p. 177). These feminist anthropologists have questioned the claim that in all societies masculinity is associated with culture and femininity with nature. Moreover, they argue that there is no behaviour or meaning which is universally and cross-culturally associated with either masculinity or femininity. What is considered masculine in some societies is considered feminine or gender-neutral in others and vice versa. Indeed, they suggest that even where the nature/culture dichotomy exists, we must not assume that the Western terms 'nature' and 'culture' are adequate or reasonable translations of the categories other cultures perceive. The historical research by Merchant and Jordanova referred to above also points to the historical specificity of these gender metaphors. As Harding says: 'the effect of these studies is to challenge the universality of the particular dichotomized set of social behaviors and meanings associated with masculinity and femininity in Western culture' (Harding, 1986, p. 129).

If we look at other cultures such as those of African and Aboriginal peoples, we find concepts of nature quite different from dominant European ones. Their world views posit a more harmonious relationship between mankind and the living universe of nature which strikingly parallels what is claimed to be a distinctively feminine world view. And what the African and Aboriginal world views designate as European is similar to what feminists designate as masculine. Even within the traditions of Western philosophy there are schools of thought which claim these values for themselves. Karl Mannheim (1953) describes romantic-conservatism as an anti-atomistic style of thinking which advocates holism, organic unity, and the qualitative rather than the quantitative as the preferred style of thought. Once more it is difficult to claim that a holistic approach in harmony with nature is specific to gender.

These arguments cast serious doubt on the projects for a feminist science presented above. Once it is recognized that 'masculinity' and 'femininity', as well as the idea of 'nature', are changing cultural categories then it no longer makes sense to base a science on feminine intuition rooted in nature. Authors like Keller, Rose and Hartsock also call for a science which incorporates women's values, although they expressly dissociate themselves from this radical feminist essentialism. Harding groups these authors under the label of the 'feminist standpoint epistemology'. This proposal argues that 'men's dominating position in social life results in partial and perverse understandings, whereas women's subjugated position provides the possibility of more complete and less perverse understandings' (Harding, 1986, p. 26). These feminist critiques of science ground a distinctive feminist science in the universal features of women's experience. Nevertheless, they all hover near the edge of biologism. Like the radical feminists, they endorse versions of a science based on subjectivity, intuition, holism and harmony. While Rose and Hartsock in particular base their materialist analyses on the gender division of labour, they fail to take fully into account that 'nature' is not a fixed category and that the division of labour is not unchanging. Therefore women's subjectivity, caring, holism and harmony, to which they appeal, cannot be universal aspects of women's experience. Their identification between women's caring labour and the new values to be incorporated into science cannot be construed as fixed or in any way as arising 'naturally'.

One attempt to overcome the limitations of the 'standpoint approach' is the critique of a feminist science from the point of view of feminist postmodernism or deconstructionism. Harding has correctly warned that the feminine qualities celebrated by feminists do not accurately reflect the social experience of all women as their experience is divided by class, race and culture. If a new feminist science is to be created from the standpoint of women's experience, should there be a feminist science based on the experience of 'Black women, Asian women, Native American women, working-class women, lesbian women?' Taking her cue from feminist postmodernism, Harding argues that the problem with feminist standpoint epistemologies is that they assume that there is a single privileged position from which science can be evaluated. There is no 'woman' to whose social experience the feminist empiricist and standpoint approaches can appeal; there are instead the 'fractured identities of women'. This approach is useful in that it takes account of the differences between and within individuals, and highlights the tension between a unitary and a fragmented conceptualization of the voice of feminism.

However the fact that there are class, race and cultural differences between women and between men does not mean that gender difference is 'either theoretically unimportant or politically irrelevant' (Harding, 1986, p. 18). In virtually every culture, gender difference is fundamental to social organization and personal identity. Qualities associated with manliness are almost everywhere more highly regarded than those thought of as womanly. Women have in common the fact that they have been marginalized from every powerful institution of our society, especially from scientific institutions. This acknowledgement of the universality of women's subordination is not incompatible with a recognition of the specific and variable forms of this subordination. Different groups of women have different needs and interests.

I share McNeil's (1987) view that rationality and intuition must themselves be seen as historically specific social products and that we should engage in social practices to redefine them. Her essay expresses well the spurious dilemma facing those feminists who feel forced to choose between scientific rationality or feminine intuition.[6] Furthermore, it is important to stress that the basis of men's power is not simply a product of the ideas we hold and the language we use, but of all the social practices that give men authority over women. Ideas are mediations of social relations and to transform them we need to transform the fundamental character of scientific institutions in contemporary society and the forms of political power that science bestows on specific social groups.

It may be that the search for the most appropriate feminist epistemology, however philosophically sophisticated (as Harding indeed is), is misdirected. The more philosophically oriented feminist work on science suffers from the problem of dealing with ideas divorced from social practices. Indeed, as amply shown by these authors, statements of 'The Scientific Method' do typically contain male visions of what it is to know and what the world is really like. Scientific practice is in no sense determined by statements of method. The latter are better seen as political pronouncements, as legitimations, rather than as descriptions of what scientists actually do. They serve to say something about the place of science in the wider society, or to bolster a more scientific speciality or discipline against its competitors (Richards and Schuster, 1989).[7]

It is in this light that we should see attempts to spell out a specifically feminist scientific method. They are politically useful in that they turn the feminist spotlight on the content of scientific knowledge instead of simply highlighting questions of recruitment to science. We need to be cautious in presuming that the adoption of a 'feminist' scientific method would lead to differences in scientific practice without a thoroughgoing change in the relations of power within science. The danger is that what might parade as feminist science would simply amount to the same scientific practice by another name.

From Science to Technology

While there has been a growing interest in the relationship of science to society over the last decade, there has been an even greater preoccupation with the relationship between technology and social change. Debate has raged over whether the 'white heat of technology' is radically transforming society and delivering us into a post-industrial age. A major concern of feminists has been the impact of new technology on women's lives, particularly on women's work. The introduction of word processors into the office provided the focus for much early research. The recognition that housework was also work, albeit unpaid, led to studies on how the increasing use of domestic technology in the home affected the time spent on housework. The exploitation of Third World women as a source of cheap labour for the manufacture of computer components has also been scrutinized. Most recently there has been a vigorous debate over developments in reproductive technology and the implications for women's control over their fertility.

Throughout these debates there has been a tension between the view that technology would liberate women - from unwanted pregnancy, from housework and from routine paid work - and the obverse view that most new technologies are destructive and oppressive to women. For example, in the early seventies, Shulamith Firestone (1970) elaborated the view that developments in birth technology held the key to women's liberation through removing from them the burden of biological motherhood. Nowadays there is much more concern with the negative implications of the new technologies, ironically most clearly reflected in the highly charged debate over the new reproductive technologies.

A key issue here is whether the problem lies in men's domination of technology, or whether the technology is in some sense inherently patriarchal. If women were in control, would they apply technology to more benign ends? In the following discussion on gender and technology, I will explore these and related questions.

An initial difficulty in considering the feminist commentary on technology arises from its failure to distinguish between science and technology. Feminist writing on science has often construed science purely as a form of knowledge, and this assumption has been carried over into much of the feminist writing on technology. However just as science includes practices and institutions, as well as knowledge, so too does technology. Indeed, it is even more clearly the case with technology because technology is primarily about the creation of artefacts. This points to the need for a different theoretical approach to the analysis of the gender relations of technology, from that being developed around science.

Perhaps this conflation of technology with science is not surprising given that the sociology of scientific knowledge over the last ten years has contested the idea of a non-controversial distinction between science and technology. John Staudenmaier (1985, pp. 83–120) comments that although the relationship between science and technology has been a major theme in science and technology studies, the discussion has been plagued by a welter of conflicting definitions of the two basic terms. The only consensus to have emerged is that the way in which the boundaries between science and technology are demarcated, and how they are related to each other, change from one historical period to another.

In recent years, however, there has been a major re-orientation of thinking about the form of the relationship between science and technology. The model of the science–technology relationship which enjoyed widespread acceptance over a long period was the traditional hierarchical model which treats technology as applied science. This view that science discovers and technology applies this knowledge in a routine uncreative way is now in steep decline. 'One thing which practically any modern study of technological innovation suffices to show is that far from applying, and hence depending upon, the culture of natural science, technologists possess their own distinct cultural resources, which provide the principal basis for their innovative activity' (Barnes and Edge, 1982, p. 149). Technologists build on, modify and extend existing technology but they do this by a creative and imaginative process. And part of the received culture technologists inherit in the course of solving their practical problems is non-verbal; nor can it be conveyed adequately by the written word. Instead it is the individual practitioner who transfers practical knowledge and competence to another. In short, the current model of the science–technology relationship characterizes science and technology as distinguishable sub-cultures in an interactive symmetrical relationship.

Leaving aside the relationship between technology and science, it is most important to recognize that the word 'technology' has at least three different layers of meaning. Firstly, 'technology' is a form of knowledge, as Staudenmaier emphasizes.[8] Technological 'things' are meaningless without the 'know-how' to use them, repair them, design them and make them. That know-how often cannot be captured in words. It is visual, even tactile, rather than simply verbal or mathematical. But it can also be systematized and taught, as in the various disciplines of engineering.

Few authors however would be content with this definition of technology as a form of knowledge. 'Technology' also refers to what people do as well as what they know. An object such as a car or a vacuum cleaner is a technology, rather than an arbitrary lump of matter, because it forms part of a set of human activities. A computer without programs and programmers is simply a useless collection of bits of metal, plastic and silicon. 'Steelmaking', say, is a technology: but this implies that the technology includes what steelworkers do, as well as the furnaces they use. So 'technology' refers to human activities and practices. And finally, at the most basic level, there is the 'hardware' definition of technology, in which it refers to sets of physical objects, for example, cars, lathes, vacuum cleaners and computers.

In practice the technologies dealt with in this book cover all three aspects, and often it is not useful to separate them further. My purpose is not to attempt to refine a definition. These different layers of meaning of 'technology' are worth bearing in mind in what follows.

The rest of this chapter will review the theoretical literature on gender and technology, which in many cases mirrors the debates about science outlined above. However, feminist perspectives on technology are more recent and much less theoretically developed than those which have been articulated in relation to science. One clear indication of this is the preponderance of edited collections which have been published in this area.[9] As with many such collections, the articles do not share a consistent approach or cover the field in a comprehensive fashion. Therefore I will be drawing out strands of argument from this literature rather than presenting the material as coherent positions in a debate.

Hidden from History

To start with, feminists have pointed out the dearth of material on women and technology, especially given the burgeoning scholarship in the field of technology studies. Even the most perceptive and humanistic works on the relationship between technology, culture and society rarely mention gender. Women's contributions have by and large been left out of technological history. Contributions to *Technology and Culture*, the leading journal of the history of technology, provide one accurate barometer of this. Joan Rothschild's (1983, pp. xii–xiv) survey of the journal for articles on the subject of women found only four in twenty-four years of publishing. In a more recent book about the journal, Staudenmaier (ibid., p. 180) also notes the extraordinary bias in the journal towards male figures and the striking absence of a women's perspective. The history of technology represents the prototype inventor as male. So, as in the history of science, an initial task of feminists has been to uncover and recover the women hidden from history who have contributed to technological developments.

There is now evidence that during the industrial era, women invented or contributed to the invention of such crucial machines as the cotton gin, the sewing machine, the small electric motor, the McCormick reaper, and the Jacquard loom (Stanley, forthcoming). This sort of historical scholarship often relies heavily on patent records to recover women's forgotten inventions. It has been noted that many women's inventions have been credited to their husbands because they actually appear in patent records in their husbands' name. This is explained in terms of women's limited property rights, as well as the general ridicule afforded women inventors at that time (Pursell, 1981; Amram, 1984; Griffiths, 1985). Interestingly, it may be that even the recovery of women inventors from patent records seriously underestimates their contribution to technological development. In a recent article on the role of patents, Christine MacLeod (1987) observes that prior to 1700 patents were not primarily about the recording of the actual inventor, but were instead sought in the name of financial backers.[10] Given this, it is even less surprising that so few women's names are to be found in patent records.

For all but a few exceptional women, creativity alone was not sufficient. In order to participate in the inventive activity of the Industrial Revolution, capital as well as ideas were necessary. It was only in 1882 that the Married Women's Property Act gave English women legal possession and control of any personal property independently of their husbands. Dot Griffiths (1985) argues that the effect of this was to virtually exclude women from participation in the world of the inventor–entrepreneur. At the same time women were being denied access to education and specifically to the theoretical grounding in mathematics and mechanics upon which so many of the inventions and innovations of the period were based. As business activities expanded and were moved out of the home, middle-class women were increasingly left to a life of enforced leisure. Soon the appropriate education for girls became 'accomplishments' such as embroidery and music – accomplishments hardly conducive to participation in the world of the inventor–entrepreneur. In the current period, there has been considerable interest in the possible contributions which Ada Lady Lovelace, Grace Hopper and other women may have made to the development of computing. Recent histories of computer programming provide substantial evidence for the view that women played a major part.[11]

To fully comprehend women's contributions to technological development, however, a more radical approach may be necessary. For a start, the traditional conception of technology too readily defines technology in terms of male activities. As I have pointed out above, the concept of technology is itself subject to historical change, and different epochs and cultures had different names for what we now think of as technology. A greater emphasis on women's activities immediately suggests that females, and in particular black women, were among the first technologists. After all, women were the main gatherers, processors and storers of plant food from earliest human times onward. It was therefore logical that they should be the ones to have invented the tools and methods involved in this work such as the digging stick, the carrying sling, the reaping knife and sickle, pestles and pounders. In this vein, Autumn Stanley (forthcoming) illustrates women's early achievements in horticulture and agriculture, such as the hoe, the scratch plow, grafting, hand pollination, and early irrigation.

If it were not for the male bias in most technology research, the significance of these inventions would be acknowledged. As Ruth Schwartz Cowan notes:

> The indices to the standard histories of technology . . . do not contain a single reference, for example, to such a significant cultural artifact as the baby bottle. Here is a simple implement . . . which has transformed a fundamental human experience for vast numbers of infants and mothers, and been one of the more controversial exports of Western technology to underdeveloped countries – yet it finds no place in our histories of technology.(1979, p. 52)

There is important work to be done not only in identifying women inventors, but also in discovering the origins and paths of development of 'women's sphere' technologies that seem often to have been considered beneath notice.

A Technology Based on Women's Values?

During the eighties, feminists have begun to focus on the gendered character of technology itself. Rather than asking how women could be more equitably treated within and by a neutral technology, many feminists now argue that Western technology itself embodies patriarchal values. This parallels the way in which the feminist critique of science evolved from asking the 'woman question' in science to asking the more radical 'science question' in feminism. Technology, like science, is seen as deeply implicated in the masculine project of the domination and control of women and nature.[12] Just as many feminists have argued for a science based on women's values, so too has there been a call for a technology based on women's values. In Joan Rothschild's (1983) preface to a collection on feminist perspectives on technology, she says that: 'Feminist analysis has sought to show how the subjective, intuitive, and irrational can and do play a key role in our science and technology'. Interestingly, she cites an important male figure in the field, Lewis Mumford, to support her case. Mumford's linking of subjective impulses, life-generating forces and a female principle is consistent with such a feminist analysis, as is his endorsement of a more holistic view of culture and technological developments.

Other male authors have also advocated a technology based on women's values. Mike Cooley is a well-known critic of the current design of technological systems and he has done much to popularize the idea of human-centred technologies. In *Architect or Bee?* (1980, p. 43) he argues that technological change has 'male values' built into it: 'the values of the White Male Warrior, admired for his strength and speed in eliminating the weak, conquering competitors and ruling over vast armies of men who obey his every instruction . . . Technological change is starved of the so-called female values such as intuition, subjectivity, tenacity and compassion'. Cooley sees it as imperative that more women become involved in science and technology to challenge and counteract the built-in male values: that we cease placing the objective above the subjective, the rational above the tacit, and the digital above analogical representation. In *The Culture of Technology*, Arnold Pacey (1983) devotes an entire chapter to 'Women and Wider Values'. He outlines three contrasting sets of values involved in the practice of technology – firstly, those stressing virtuosity, secondly, economic values and thirdly, user or need-oriented values. Women exemplify this third 'responsible' orientation, according to Pacey, as they work with nature in contrast to the male interest in construction and the conquest of nature.

Ironically the approach of these male authors is in some respects rather similar to the eco-feminism that became popular amongst feminists in the eighties. This marriage of ecology and feminism rests on the 'female principle', the notion that women are closer to nature than men and that the technologies men have created are based on the domination of nature in the same way that they seek to dominate women. Eco-feminists concentrated on military technology and the ecological effects of other modern technologies. According to them, these technologies are products of a patriarchal culture that 'speaks violence at every level' (Rothschild, 1983, p. 126). An early slogan of the feminist anti-militarist movement, 'Take the Toys from the Boys', drew attention to the phallic symbolism in the shape of missiles. However, an inevitable corollary of this stance seemed to be the representation of women as inherently nurturing and pacifist. The problems with this position have been outlined above in relation to science based on women's essential values. We need to ask how women became associated with these values. The answer involves examining the way in which the traditional division of labour between women and men has generally restricted women to a narrow range of experience concerned primarily with the private world of the home and family.

Nevertheless, the strength of these arguments is that they go beyond the usual conception of the problem as being women's exclusion from the processes of innovation and from the acquisition of technical skills. Feminists have pointed to all sorts of barriers - in social attitudes, girls' education and the employment policies of firms - to account for the imbalance in the number of women in engineering. But rarely has the problem been identified as the way engineering has been conceived and taught. In particular, the failure of liberal and equal opportunity policies has led authors such as Cynthia Cockburn (1985) to ask whether women actively resist entering technology. Why have the women's training initiatives designed to break men's monopoly of the building trades, engineering and information technology not been more successful? Although schemes to channel women into technical trades have been small-scale, it is hard to escape the conclusion that women's response has been tentative and perhaps ambivalent.

I share Cockburn's view that this reluctance 'to enter' is to do with the sex-stereotyped definition of technology as an activity appropriate for men. As with science, the very language of technology, its symbolism, is masculine. It is not simply a question of acquiring skills, because these skills are embedded in a culture of masculinity that is largely coterminous with the culture of technology. Both at school and in the workplace this culture is incompatible with femininity. Therefore, to enter this world, to learn its language, women have first to forsake their femininity.

Technology and the Division of Labour

I will now turn to a more historical and sociological approach to the analysis of gender and technology. This approach has built on some theoretical foundations provided by contributors to the labour process debate of the 1970s. Just as the radical science movement had sought to expose the class character of science, these writers attempted to extend the class analysis to technology. In doing so, they were countering the theory of 'technological determinism' that remains so widespread.

According to this account, changes in technology are the most important cause of social change. Technologies themselves are neutral and impinge on society from the outside; the scientists and technicians who produce new technologies are seen to be independent of their social location and above sectional interests. Labour process analysts were especially critical of a technicist version of Marxism in which the development of technology and productivity is seen as the motor force of history. This interpretation represented technology itself as beyond class struggle.

With the publication of Harry Braverman's *Labor and Monopoly Capital* (1974), there was a revival of interest in Marx's contribution to the study of technology, particularly in relation to work. Braverman restored Marx's critique of technology and the division of labour to the centre of his analysis of the process of capitalist development. The basic argument of the labour process literature which developed was that capitalist–worker relations are a major factor affecting the technology of production within capitalism. Historical case studies of the evolution and introduction of particular technologies documented the way in which they were deliberately designed to deskill and eliminate human labour.[13] Rather than technical inventions developing inexorably, machinery was used by the owners and managers of capital as an important weapon in the battle for control over production. So, like science, technology was understood to be the result of capitalist social relations.

This analysis provided a timely challenge to the notion of technological determinism and, in its focus on the capitalist division of labour, it paved the way for the development of a more sophisticated analysis of gender relations and technology. However, the labour process approach was gender-blind because it interpreted the social relations of technology in exclusively class terms. Yet, as has been well established by the socialist feminist current in this debate, the relations of production are constructed as much out of gender divisions as class divisions. Recent writings (Cockburn, 1983, 1985; Faulkner and Arnold, 1985; McNeil, 1987) in this historical vein see women's exclusion from technology as a consequence of the gender division of labour and the male domination of skilled trades that developed under capitalism. In fact, some argue that prior to the industrial revolution women had more opportunities to acquire technical skills, and that capitalist technology has become more masculine than previous technologies.

I have already described how, in the early phases of industrialization, women were denied access to ownership of capital and access to education. Shifting the focus, these authors show that the rigid pattern of gender divisions which developed within the working-class in the context of the new industries laid the foundation for the male dominance of technology. It was during this period that manufacturing moved into factories, and home became separated from paid work. The advent of powered machinery fundamentally challenged traditional craft skills because tools were literally taken out of the hands of workers and combined into machines. But as it had been men who on the whole had technical skills in the period before the industrial revolution, they were in a unique position to maintain a monopoly over the new skills created by the introduction of machines.

Male craft workers could not prevent employers from drawing women into the new spheres of production. So instead they organized to retain certain rights over technology by actively resisting the entry of women to their trades. Women who became industrial labourers found themselves working in what were considered to be unskilled jobs for the lowest pay. 'It is the most damning indictment of skilled working-class men and their unions that they excluded women from membership and prevented them gaining competences that could have secured them a decent living' (Cockburn, 1985, p. 39). This gender division of labour within the factory meant that the machinery was designed by men with men in mind, either by the capitalist inventor or by skilled craftsmen. Industrial technology from its origins thus reflects male power as well as capitalist domination.

The masculine culture of technology is fundamental to the way in which the gender division of labour is still being reproduced today. By securing control of key technologies, men are denying women the practical experience upon which inventiveness depends. I noted earlier the degree to which technical knowledge involves tacit, intuitive knowledge and 'learning by doing'. New technology typically emerges not from sudden flashes of inspiration but from existing technology, by a process of gradual modification to, and new combinations of, that existing technology. Innovation is to some extent an imaginative process, but that imagination lies largely in seeing ways in which existing devices can be improved, and in extending the scope of techniques successful in one area into new areas. Therefore giving women access to formal technical knowledge alone does not provide the resources necessary for invention. Experience of existing technology is a precondition for the invention of new technology.

The nature of women's inventions, like that of men's, is a function of time, place and resources. Segregated at work and primarily confined to the private sphere of the household, women's experience has been severely restricted and therefore so too has their inventiveness. An interesting illustration of this point lies in the fact that women who were employed in the munitions factories during the First World War are on record as having redesigned the weaponry they were making.[14] Thus, given the opportunity, women have demonstrated their inventive capacity in what now seems the most unlikely of contexts.

Missing: The Gender Dimension in the Sociology of Technology

The historical approach is an advance over essentialist positions which seek to base a new technology on women's innate values. Women's profound alienation from technology is accounted for in terms of the historical and cultural construction of technology as masculine. I believe that women's exclusion from, and rejection of, technology is made more explicable by an analysis of technology as a culture that expresses and consolidates relations amongst men. If technical competence is an integral part of masculine gender identity, why should women be expected to aspire to it?

Such an account of technology and gender relations, however, is still at a general level.[15] There are few cases where feminists have really got inside the 'black box' of technology to do detailed empirical research, as some of the most recent sociological literature has attempted. Over the last few years, a new sociology of technology has emerged which is studying the invention, development, stabilization and diffusion of specific artefacts.[16] It is evident from this research that technology is not simply the product of rational technical imperatives. Rather, political choices are embedded in the very design and selection of technology.

Technologies result from a series of specific decisions made by particular groups of people in particular places at particular times for their own purposes. As such, technologies bear the imprint of the people and social context in which they developed. David Noble (1984, p. xiii) expresses this point succinctly as follows: 'Because of its very concreteness, people tend to confront technology as an irreducible brute fact, a given, a first cause, rather than as hardened history, frozen fragments of human and social endeavor'. Technological change is a process subject to struggles for control by different groups. As such, the outcomes depend primarily on the distribution of power and resources within society.

There is now an extensive literature on the history of technology and the economics of technological innovation. Labour historians and sociologists have investigated the relationship between social change and the shaping of production processes in great detail and have also been concerned with the influence of technological form upon social relations. The sociological approach has moved away from studying the individual inventor and from the notion that technological innovation is a result of some inner technical logic. Rather, it attempts to show the effects of social relations on technology that range from fostering or inhibiting particular technologies, through influencing the choice between competing paths of technical development, to affecting the precise design characteristics of particular artefacts. Technological innovation now requires major investment and has become a collective, institutionalized process. The evolution of a technology is thus the function of a complex set of technical, social, economic, and political factors. An artefact may be looked on as the 'congealed outcome of a set of negotiations, compromises, conflicts, controversies and deals that were put together between opponents in rooms filled with smoke, lathes or computer terminals' (Law, 1987, p. 406).

Because social groups have different interests and resources, the development process brings out conflicts between different views of the technical requirements of the device. Accordingly, the stability and form of artefacts depends on the capacity and resources that the salient social groups can mobilize in the course of the development process. Thus in the technology of production, economic and social class interests often lie behind the development and adoption of devices. In the case of military technology, the operation of bureaucratic and organizational interests of state decision-making will be identifiable. Growing attention is now being given to the extent to which the state sponsorship of military technology shapes civilian technology.

So far, however, little attention has been paid to the way in which technological objects may be shaped by the operation of gender interests. This blindness to gender issues is also indicative of a general problem with the methodology adopted by the new sociology of technology. Using a conventional notion of technology, these writers study the social groups which actively seek to influence the form and direction of technological design. What they overlook is the fact that the absence of influence from certain groups may also be significant. For them, women's absence from observable conflict does not indicate that gender interests are being mobilized. For a social theory of gender, however, the almost complete exclusion of women from the technological community points to the need to take account of the underlying structure of gender relations. Preferences for different technologies are shaped by a set of social arrangements that reflect men's power in the wider society. The process of technological development is socially structured and culturally patterned by various social interests that lie outside the immediate context of technological innovation.

More than ever before technological change impinges on every aspect of our public and private lives, from the artificially cultivated food that we eat to the increasingly sophisticated forms of communication we use. Yet, in common with the labour process debate, the sociology of technology has concentrated almost exclusively on the relations of paid production, focusing in particular on the early stages of product development. In doing so they have ignored the spheres of reproduction, consumption and the unpaid production that takes place in the home. By contrast, feminist analysis points us beyond the factory gates to see that technology is just as centrally involved in these spheres.

Inevitably perhaps, feminist work in this area has so far raised as many questions as it has answered. Is technology valued because it is associated with masculinity or is masculinity valued because of the association with technology? How do we avoid the tautology that 'technology is masculine because men do it'? Why is women's work undervalued? Is there such a thing as women's knowledge? Is it different from 'feminine intuition'? Can technology be reconstructed around women's interests? These are the questions that abstract analysis has so far failed to answer. The character of salient interests and social groups will differ depending on the particular empirical sites of technology being considered. Thus we need to look in more concrete and historical detail at how, in specific areas of work and personal life, gender relations influence the technological enterprise. This book focuses on gender, although it is often difficult to disentangle the effects of gender from those of class and race. The chapters that follow are organized around substantive areas of technology – the technology of production, reproductive technology, domestic technology and the built environment.

Throughout the book I will be stressing that a gendered approach to technology cannot be reduced to a view which treats technology as a set of neutral artefacts manipulated by men in their own interests. While it is the case that men dominate the scientific and technical institutions, it is perfectly plausible that there will come a time when women are more fully represented in these institutions without transforming the direction of technological development. To cite just one instance, women are increasingly being recruited into the American space-defence programme but we do not hear their voices protesting about its preoccupations. Nevertheless, gender relations are an integral constituent of the social organization of these institutions and their projects. It is impossible to divorce the gender relations which are expressed in, and shape technologies from, the wider social structures that create and maintain them. In developing a theory of the gendered character of technology, we are inevitably in danger of either adopting an essentialist position that sees technology as inherently patriarchal, or losing sight of the structure of gender relations through an overemphasis on the historical variability of the categories of 'women' and 'technology'. In what follows I will try to chart another course.

NOTES

1 For an introduction to this literature, see Barnes and Edge (1982) and Knorr-Cetina and Mulkay (1983).
2 In order to map the field of gender and science, I have drawn heavily on two excellent and comprehensive surveys by Harding (1986) and Schiebinger (1987).
3 This issue is discussed in Harding (1986). For a fuller account of the debate about whether Reason itself is male, see Lloyd (1984).
4 For an excellent discussion of Keller's work, see Dugdale (1988).
5 For two useful socialist feminist critiques of universalist and essentialist elements in some versions of radical feminist theory, see Eisenstein (1984) and Segal (1987).
6 For an account of the way the binary couple 'empiricism-inductivism'/ 'intuitive-speculative theory building' has been played upon since the seventeenth century, see Schuster and Yeo (1986).
7 For a clever comparison of the biographies of McClintock and Franklin and their respective scientific methodologies, see Richards and Schuster (1989).
8 Staudenmaier (1985, pp. 103–20) outlines four characteristics of technological knowledge–scientific concepts, problematic data, engineering theory, and technological skill.

9 A good cross-section of this material can be found in Trescott (1979); Rothschild (1983); Faulkner and Arnold (1985); McNeil (1987); Kramarae (1988). McNeil's book is particularly useful as it contains a comprehensive bibliography which is organized thematically.

10 MacLeod (1987) suggests that although George Ravenscroft is credited in the patent records with being the 'heroic' inventor of lead-crystal glass, he was rather the purchaser or financier of another's invention. This study alerts us to the danger of assuming that patent records have always represented the same thing.

11 For a biography of Lady Lovelace, which takes issue with the view of her as a major contributor to computer programming, see Stein (1985). However, both Kraft (1977) and more recently Giordano (1988) have documented the extensive participation of women in the development of computer programming.

12 Technology as the domination of nature is also a central theme in the work of critical theorists, such as Marcuse, for whom it is capitalist relations (rather than patriarchal relations) which are built into the very structure of technology. 'Not only the application of technology but technology itself is domination (of nature and men) - methodical, scientific, calculated, calculating control. Specific purposes and interests of domination are not foisted upon technology 'subsequently' and from the outside; they enter the very construction of the technical apparatus' (Marcuse, 1968, pp. 223–4).

13 This point is elaborated in the next chapter. See also Part Two of MacKenzie and Wajcman (1985) for a collection of these case studies.

14 Amram (1984) provides a selection of the patents granted to women during the First World War.

15 Cockburn's (1983, 1985) work is one important exception discussed at greater length in chapter 2.

16 For an introduction to this literature, see MacKenzie and Wajcman (1985); Bijker, Hughes and Pinch (1987).

2
The Technology of Production: Making a Job of Gender

> Capitalists as capitalists and men as men both take initiatives over technology. Cockburn, *The Material of Male Power*

Our images of technology are starkest in the sphere of production and paid work – from dark satanic mills to clean, automated factories run almost entirely by robots. After all, people depend for their livelihood on paid work and it is here that they spend most of their time. This is where some of the fiercest battles over the costs and benefits of technological change have been fought. The most notorious involved male weavers in nineteenth-century England destroying the new machines and mills that threatened their jobs. Indeed, the term 'Luddite' is still used to denote those who oppose technological change and thus stand in the way of progress.

The late twentieth century finds us in another period of rapid technological development. Fundamental innovations in microelectronic and telecommunications technology are transforming the character of work and the structure of the labour force. Existing sexual divisions of labour are profoundly implicated in these changes, and new terrain for the gendering of work is being opened up. These are the issues which I will consider in this chapter.

Whether new technology is a liberating force which will eliminate the dehumanizing aspects of work or whether it will inevitably lead to the degradation, fragmentation, and intensification of technologically-paced work depends on your point of view. Theorists of post-industrial society such as Alvin Toffler (1980), Barry Jones (1982) and André Gorz (1982) are optimistic about the radical changes that they believe are emerging in industrial societies as a result of the 'microelectronic revolution'. They argue that technological innovations mean less labour being expended on industrial development, and a shift from manufacturing to service-based economies. The jobs destroyed by microelectronics in industry would be replaced by new occupations in these new industries, and degraded and routine work would be consigned to machines, releasing human beings to more creative and more fulfilling work. In the future some of this work may even shift to the home or 'electronic cottage', as computerization will eliminate the need for people to work in large-scale units of production.

Another version of this optimistic view has recently emerged as the sociology of work has become increasingly concerned with the issue of 'flexible specialization' and 'neo-Fordism'. The focus here tends to be on the potential for job enhancement presented by new technology. To put it simply, automation will increase skill requirements. It is argued that technological change, especially in conjunction with the use of Japanese management techniques, will require a smaller, but skilled and flexible workforce, prepared to undertake ongoing training to facilitate the adaptation of skills to new technology.[1]

Rising unemployment levels in advanced capitalist societies have prompted a more pessimistic view of technology's impact on work. In contrast to the 'post-industrial' scenario, such commentators believe that automation is associated with degraded, deskilled, and devalued jobs, stressful and dangerous work, employer monitoring of employees, and work speed-ups, in which workers are paid less for doing more. As workers' skills are built into the technology, those fortunate enough to retain a job are relegated to the position of machine minders. (This scenario is vividly depicted in Kurt Vonnegut's science fiction novel *Player Piano*.) The increasing use by management of surveillance systems built into the machinery itself to monitor and record output will serve to intensify the exploitation of labour.

Conducted in these terms, the debate begs many questions. Is the technology of production an independent force determining the organization of work? In particular, does it affect the nature and experience of work for women and men alike? Or is the development and introduction of particular technologies itself shaped by existing social relations, including those of gender?

Although new technologies do represent a force for change, I will be arguing that the outcomes are constrained by the pre-existing organization of work, of which gender is an integral part. Technical change has not substantially undermined sexual divisions in the labour market and occupational segregation between women and men. This raises the question of what has shaped particular technological developments in the first place. If technology is designed with job stereotypes in mind then it is hardly surprising that sex segregation is being further incorporated into the workplace. Accordingly, the chapter will go on to explore the ways in which the sex of the workforce and gender relations in the workplace themselves profoundly affect the direction and pace of technological innovation.

The Impact of Technology on the Sexual Division of Labour

Office Automation and Women's Employment

Whilst women have always worked in large numbers, it is over the last three decades that they have come to comprise nearly half the labour force in advanced industrial economies. Even so, many of the protagonists in the debate on work and technology have been oblivious to gender issues, implicitly concerned only with those sectors of production in which male workers predominate. Since the mid-1970s, however, feminist researchers and activists have addressed the effects of automation on women's employment.

The introduction of computer-based technologies into offices has been the focus of one strand of this research, mainly because the majority of clerical and secretarial workers almost everywhere are women. It is also the case that these groups are being disproportionately affected, as the office is the prime site of technologically induced change. This research forms the basis for many of the generalizations about women's work experience.

Optimistic and pessimistic views can be discerned in the various studies on office automation.[2] Some authors suggest that the introduction of word processing equipment is making the traditional secretarial job obsolete. But, rather than being deskilled, they see the job of secretary as being replaced by different types of paraprofessional jobs. Routine typing would be minimized leaving the office worker to take on more skilled, satisfying work as well as more responsible duties. Technological advances will improve the quality of work, reducing drudgery and promoting more integrated work processes. This optimistic vision attaches great significance to the liberating potential of new office technologies, seeing in them a solution to women's traditionally limited and limiting work opportunities.

Much more common among feminist writers however has been a pessimistic view of the impact of microelectronic technology on women's work, often expressed in a strongly anti-technology stance. A major concern in the women's movement has been the implications for women's health and safety of widespread use of video display terminals, from eye strain and headaches to the risks of radiation for pregnant women. Many surveys of users have reported physical and psychological symptoms, such as vision problems, tenosynovitis or repetition strain injury, chest pain, tension, headaches, nausea and dizziness, digestive problems, and depression. This is particularly so for those who were subject to computerized work monitoring suggesting that the intensity of work is a major cause of these stress-related illnesses. Setting strict limits on the time spent at terminals has therefore been a major international issue in trade union negotiations over new technology.

More generally, fears have been expressed that computerization of office work would lead to a huge reduction in the number of office jobs and the emergence of the 'paperless office'. Word processors were seen as a threat to typists' skills which were being incorporated into the new machines. Secretarial work for those few who remained would be increasingly fragmented into routine and standardized tasks subject to the control of the machine.

To understand the genesis of this negative position we need to look at the framework in which the debate has developed. The terms of the feminist discussion have been influenced by what is known as 'the labour process perspective' or the deskilling debate, discussed in chapter 1.[3] Labour process theorists have criticized technological determinism, arguing that, far from constituting an autonomous force determining the organization of work, technology is itself crucially affected by the antagonistic class relations of production. According to this view, capitalism requires the continuous application of new technology to the fragmentation and cheapening of labour, resulting in deskilling.

The introduction of information technologies into the office has been seen as part of the general process of deskilling. The purpose is to increase productivity and profit, in this case by deskilling typists and incorporating the monitoring of work into the machinery itself. Labour process analysis characterizes the office as a white-collar replica of the assembly line, with office work broken into many sub-tasks, each performed by a specialized worker, who loses both contact with the total product and variety in the tasks performed. With this rationalization of the office, the conditions of white-collar work become increasingly like factory work. Hence the (well-known) term, the 'proletarianization' of white-collar workers. Through this process, management reduces the skill requirements of office work and thus reduces labour costs. The result is that workers have less and less control over the production process.

However, reality is more complex than the proletarianization thesis suggests, as detailed empirical studies of technological change have repeatedly shown. In particular, they have questioned the existence of any simple tendency towards either the deskilling or the upgrading of labour. Indeed, with respect to the skill levels required for given jobs, opposing tendencies of increased complexity and of greater simplification and standardization have coexisted. Some authors assumed that the machinery itself had some inherent logic which caused its impact to be uniform across the range of office jobs. In fact, identical equipment, in this case word processors, may have very different effects on work experience.

Any analysis of office automation must consider the different positions of office workers within the white-collar hierarchy, the degree of fragmentation of office work before the introduction of word processors, and the requirements of particular employers at particular periods. Although the effects of particular technologies must vary in different contexts, it has become clear that the overall tendency is for technology-led changes to operate within and reinforce preexisting differences in the patterns of work. Technological change thus tends to further advantage those who already have recognized skills and a degree of control over their work tasks.[4]

The effects of new technology on typists and on secretaries in Britain are a case in point. Juliet Webster's (1989) comparative study found that rather than automation transforming these occupations it entrenched the inequalities between them. The rationalization and fragmentation of clerical work had in fact long predated the advent of computer technology and its introduction reinforced this tendency for typists to perform repetitive, standardized tasks.[5] At the same time, however, word processors reduced the burden of routine work for the secretaries, enabling them to continue to do a variety of relatively responsible tasks. Thus the introduction of word processors exacerbated preexisting divisions between two groups of women office workers, enhancing the position of some secretaries but not that of typists.

The contradictory picture that emerges from attempts to develop general theories of the evolution of office work is partly a result of the fact that inappropriate comparisons are made between experiences at different stages in the evolution of technologies. Given rapid changes in the technology itself, in its uses, and in the cumulation of its effects, conclusions from one wave of technology may not generalize to later waves. An examination of the technological transformation of the American insurance industry is instructive here.

During the first wave of electronic automation, the computerization of some aspects of underwriting and rating occurred without major reorganization of production, leading to job loss amongst traditional female clerical workers. Barbara Baran (1987) argues that it was in the late 1970s that the insurance industry was radically restructured and that this was accompanied by the feminization of that labour force. While the early use of computers was based on, and reinforced, the fragmentation of jobs within established hierarchies, the newer applications of information technology integrated fragmented tasks to create new jobs, while often eliminating old ones. By 1983, a new highly skilled clerical position had been designed for women with college degrees. In this industry, automation had resulted in the deskilling of male professional functions with female professionals earning considerably less than the men previously did. However, although skilled clerical positions had expanded, there was little opportunity for career advancement. Furthermore, the increased emphasis on college education, with the loss of unskilled clerical jobs, was likely to narrow opportunities for less educated black and white urban working-class women.

This study shows how important it is to periodize the process of technological change. It also points to the different effects automation may have on different groups of women workers at different times.

As I said above, many of these studies of office automation were heavily influenced by labour process theory. They concentrated on the way in which capitalist management used new technologies to deskill and subordinate workers. The deskilling of craft workers was, and largely still is, the central issue in this analysis of technical change. Drawing on these studies of deskilling in male crafts, the studies on office work tended to romanticize the typist's job before the introduction of word processors by depicting it as a combination of technical craft-type skills with control over the labour process.

Women's office work is not akin to craft work. Craft workers were an elite group who enjoyed a privileged position in the labour market and considerable autonomy over the labour process. Not only is this romanticized version wrong in the specific instance of office work, but it is wrong in general terms. Women have traditionally been excluded from craft labour. An analysis based on the loss of craft skills is thus not a relevant one for women.

The more substantive problem reflected in these early studies was the assumption that the social relations within which technology developed could be understood simply in terms of relations between worker and capitalist. This underestimated the continuing significance of divisions within the working class, such as those based on sex, race, age and skill, in shaping the effects of technical change on the workplace. Feminist writers have been important in shifting the focus from a principal concern with class conflict. In particular, they have exposed the inappropriateness of the craft model by highlighting the exclusivity of craft unions as male preserves. Craft unions have played an active part in creating and sustaining women's subordinate position in the workforce. Any understanding of technology will be incomplete without the recognition that the relations of production are constructed as much out of gender divisions as class divisions.

New Technology and Gender Relations

The more recent work on gender and technology goes beyond looking at women workers as such. Rather it has taken up this issue of divisions between workers and focused on the relationship between men and women in the workplace, the implications for the construction of jobs and the sex-typing of occupations. This has been part of the growing recognition of the limitations of sociological accounts which analyse women's position in the labour force primarily with reference to the domestic division of labour.

This is not to deny that women's disadvantaged position in the labour market is in large part due to their greater responsibility for dependent care and household tasks, as I have explored elsewhere (Wajcman, 1983). However gender relations are embodied in the sphere of production, as well as in the sphere of reproduction. Thus the gender stereotyping of jobs is not just a reflection of women's traditional role within the family; it is also formed and reproduced by the patriarchal relations of paid work.

Some commentators have presumed that with technological developments, such as the elimination of much heavy physical work by mechanization, the boundaries between women's and men's work would gradually disappear. From a different perspective, the labour process literature presumed that women would become fully integrated into the labour force as technology led to its increasing homogenization (Liff, 1986, p. 75).

However, the gender stereotyping of jobs has remained remarkably stable even when the nature of work and the skills required to perform it have been radically transformed. The broad nature of gender divisions in the labour market is well established: men and women are segregated into different occupations, and this segregation is particularly marked within individual workplaces. Women are, on average, paid about three-quarters of men's hourly earnings. What requires explanation is the contrast between the flexibility of the form taken by occupational segregation by sex and its persistence.

It has been more common for women to enter new jobs requiring new skills than to break into traditional male preserves, as the example of the insurance industry shows. Even the allocation of these completely new jobs, where no gendered custom and practice has been established, is a fundamentally gendered process. In new 'high-tech' jobs, such as programming, women tend to be segregated into positions at the bottom of the occupational hierarchy (Kraft and Dubnoff, 1986). Much of the recent feminist work has addressed the issue of why there has been so little change in the degree of sex segregation of the labour market and so little conflict over the continuing rigid sex-typing of occupations.[6]

What role does technology play in the construction and reproduction of these gender relations, and in their potential transformation? New technologies do disrupt established patterns of sex-typing and thereby open up opportunities for changing the sexual division of labour. As technologies develop and displace each other, there is a disturbance among the technically skilled strata. Some gain and some lose. Many male craft skills have been quite purposively made redundant by new technology that has radically transformed the nature of the work. But technology is not an independent force; the way in which it affects the nature of work is conditioned by existing relationships. There are conflicts and negotiations over technological change and the opportunities for changing the sexual division of labour to women's advantage are often foreclosed by male power. Women lose out in these struggles as powerful groups defend their old skills or monopolize new ones. Craft workers, who have been seen as the defenders of working-class interests in struggles over technical change, in part derive their strength from their past exclusionary practices. Their gains have often been made at the expense of less skilled or less well-organized sections of the workforce, and this has in many cases involved the exclusion of women.

The entry of women into industrial work in Britain, America and Australia during the First and, especially, the Second World Wars was followed by an equally deliberate process of their expulsion from that work once the immediate crisis had passed (see figure 2.1). Thus the gross under-representation of women in engineering and other industrial work, and the lack of confidence often felt by women faced with technology, are evidence of a deeper problem. Official plans to rectify the under-representation of women in engineering often proceed as

2.1 Munitions worker during World War II
Source: Australian War Memorial, negative no. 13178. From the film *For Love or Money* (1983) by Megan McMurchy, Margot Oliver, Jenni Thornley

if the problem were simply a lack of self-confidence in women. But male dominance of technology has in large part been secured by the active exclusion of women from areas of technological work.

Printing and newspaper publishing in particular is an industry with craft traditions of labour process control. Recent technological developments, particularly in electronic typesetting technologies which have the potential to undermine those traditions, have been resisted by printing workers. Strikes and lockouts throughout the 1970s and early 1980s have characterized attempts to introduce the new technology in the United States, Great Britain and Australia. The printing industry in Britain provides a contemporary illustration of the sexual politics involved in such struggles over technology.

The violent dispute at the new technology newspaper plant at

Wapping in London during 1986 was the final phase in a long history of management attempts to wrest control of the labour process from the Fleet Street print unions. Computerized photocomposition systems had been available in the United States printing industry from about 1970. This technology enabled journalists and advertising personnel to enter copy directly into a computer. The introduction of this new technology represented an attack on the compositors' control over their work as it meant that their traditional manual skills would become technically redundant.

The restrictive practices of craft labour and the degree of chapel control had, not surprisingly, inhibited the introduction of this kind of technological innovation into the newspaper industry in London's Fleet Street. Wages of these craftsmen have been very high in the post-war period. Those of compositors, who prepared the type in hot metal, were the highest of all. In her book on the history of typesetting technology in Britain, Cynthia Cockburn (1983) describes this archetypal group of skilled male workers as they were being radically undermined by cold electronic composition. This is an area of work from which women have been traditionally excluded. Employers saw this technical change as enabling them to replace the men with cheaper women workers. Over the past decade the compositors have fought to defend their position by having sole rights to use the computer typesetting equipment - to retain keyboard work. To varying degrees they managed to maintain their craft control even though their craft was technologically redundant. However, their strategy of resistance has entailed the exclusion of unskilled women from the trade. It should be noted that this exclusionary strategy has also involved racial and religious prejudice. Skilled printing workers have a higher proportion of white Anglo-Saxon Protestants among them than the semi- and unskilled.

Having pointed out the way in which organized male workers have used technology to maintain power over women in the workplace, it needs to be said that this is not a once-and-for-all achievement. Male dominance over machinery is constantly under threat - both by women's direct efforts to undermine it, and by actions of employers in seeking to undermine skilled male workers and cheapen their labour costs.

Under some conditions, skilled men do lose out and women enter previously male jobs. The process of feminization is often part of technological change. In such cases, women rarely perform exactly the same tasks, under the same conditions, as the men formerly performed: inherent in this process of technological change is the transformation of jobs. However, and this is the crucial point, the introduction of female labour is usually accompanied by a downgrading of the skill content of the work and a consequent fall in pay for the job.

Sex, Skill and Technical Competence

It is often said that women are low paid because they are unskilled; certainly women's work tends to fall into the unskilled or semi-skilled categories of official classifications. But the crucial question is how definitions of skill are established. To take a simple example, women who assemble digital watches and pocket calculators require considerable manual dexterity ('nimble fingers'), the capacity for sustained attention to detail and excellent hand–eye co-ordination. Yet these capacities are not defined as 'skills'. Nurses provide another example of an occupation that requires a great deal of training and ability, as well as technical knowledge. However nursing is not thought of as a technical job because it is women's work. Moreover, because such work has been socially constructed as unskilled it has also been undervalued. Consequently 'women's work' is comparatively low paid. The work of women is often deemed inferior simply because it is women who do it.

How has it come about that women have failed to achieve recognition of the skills required by their work? Although it is the case that women workers have generally been refused access to training in traditionally masculine areas of work, the basis for distinctions of skill in women's and men's work is not a simple technical matter. Definitions of skill can have more to do with ideological and social constructions than with technical competencies which are possessed by men and not by women. It is a question of workers' collective efforts to protect and secure their conditions of employment – by retaining skill designations for their own work and defending that skill to the exclusion of outsiders. These efforts have been predominantly by and on behalf of the male working class. They have been directed against employers who have regularly tried to find ways of substituting cheaper workers for expensive skilled labour.

But men's resistance has also operated against women's interests. Defending skill, preventing 'dilution', has almost always meant blocking women's access to an occupation. Moreover employers' own perceptions of the suitability of women for particular types of work must in part be responsible for the craft workers' success in excluding women from skilled work (Liff, 1986). Otherwise one would expect the sexual division of labour to be a much more contested area both for management and unions than it is. Skilled status has thus been traditionally identified with masculinity and as work that women don't do, while women's skills have been defined as non-technical and undervalued.

Thus there are important connections between men's power in the workplace and their dominance over machinery. Likewise, there are important connections between women's relative lack of power and their lack of technological skills. In chapter 1 I said that technology includes not just things themselves but the physical and mental know-how to make use of these things. Know-how is a resource that gives those who possess it a degree of actual or potential power and we have seen above how this know-how has been central to the class politics of technological work. It is also central to the sexual politics of technological work, as technical competence is a key source of men's power over women – of the capacity, for example, to command higher incomes and scarce jobs.

How can we begin to understand the enduring force of this identification between technical skills and masculinity without making the mistake of treating technology as inherently masculine? We can start, as Cockburn does, by taking seriously the requirement to understand the masculinity of technology as a social product. Men's affinity with technology is then seen as integral to the constitution of male gender identity. 'Technology enters into our sexual identity: femininity is incompatible with technological competence; to feel technically competent is to feel manly' (Cockburn, 1985, p. 12).

Once we recognize that gender construction is an ongoing ideological and cultural process with a long history, then the focus shifts to analyzing the social practices involved. The way in which the present technical culture expresses and consolidates relations among men becomes an important factor in explaining the continuing exclusion of women.

This type of analysis stresses the importance of the cultural aspects of gender relations, and shows the way that gender is an integral part of people's experience in the workplace. This is illustrated in Cockburn's (1983) study of compositors, where she ascribes the centrality of the craft workers' ease with technology to their masculine identity. The industrial strength of craftsmen derived from their knowledge and competence with machines. The control over this type of industrial technology has traditionally been the province of men, and women workers have been excluded from these technical skills. The technical change from linotype to electronic photocomposition, however, literally makes the compositors feel emasculated. Because the work of composing now resembles typing and involves working with paper instead of metal, that is, a shift from factory work to the office, the compositors no longer consider it to be real work. Traditional craft culture was associated with hot metal, dirt and physical work and the elimination of this not only diminishes their control over their work but it also represents a threat to their masculinity.

Clearly, however, the appropriation of technical skills plays an important part in the reproduction of inequality among men as well as between men and women. Men do not have power over women in the same sense as capitalists do over workers. In looking at the relations of work one is inevitably looking at class relations. The male culture of craft know-how is the culture of an exploited group. Male employees themselves vary considerably in their capacities to control and benefit from technological innovations. It is important to remember that this source of power is a subordinate one in that technology is also used by some men to dominate others.

The class dimension is also significant in another sense. It is not the case that all women have an identical relationship to machinery and to technical knowledge. There are obviously important differences between the technical skills of say women factory workers and those of technically trained professional women. However, Cockburn found that what they had in common was that they were both to be found operating machinery, but rarely in those occupations that involve knowing what goes on inside the machine. 'With few exceptions, the designer and developer of the new systems, the people who market and sell, install, manage and service machinery, are men. Women may push the buttons but they may not meddle with the works' (1985, pp. 11–12). Women may well have considerable knowledge about the machine that they work on, but the key to power is flexible, transferable skills and these are still the property of men.

To say that technical competence is part of male gender identity, is not to presume that there is a coherent single form of masculinity. The masculine culture of technology may take a partially different form for working-class and middle-class men. The cult of masculinity revolving around physical prowess is closely associated with shop-floor culture among manual workers. Working-class men may be more able when it comes to fixing cars and domestic machines, but middle-class men have more power through their possession of abstract and generalizable technical knowledge. Furthermore, it needs to be stressed that ethnic and generational differences, as well as class divisions, produce different versions of masculinity. If we are to avoid essentialist constructions of 'men' and 'masculinity', we need to pluralize the term and speak of 'masculinities'. I will return to a more extensive discussion of the nature of masculine technological culture in chapter 6.

The Relocation of Work

It has been widely noted that the development of microelectronic and telecommunication technology opens up the possibility of radical changes in the location of work. White-collar work, for example, can be decentralized and moved into suburban offices (with lower rents and possibly lower wages) or it can be moved 'offshore' altogether. Sending work offshore, while not new, is certainly much easier as a result of greater satellite telecommunications capacity. An international sexual division of labour has emerged based on the breakdown of the production process in computer manufacture, with women performing the labour-intensive assembly of microchips in various Third World countries. More recently, offshore office services have developed where low-wage female labour is used for data entry and data processing work for firms based in the industrialized countries. It should be noted, however, that just as in manufacturing the development of advanced automation systems has reduced the need for offshore assembly work, so developments in office automation (such as voice recognition and optical character recognition) suggest that the use of offshore office services will be a short-term phenomenon.

The development of computer-based homework, which is also referred to as 'telework' or 'telecommuting', illustrates further the impact that technology has on the location of work. The combination of computer and telecommunications technology has made it technically feasible for large numbers of workers whose jobs involve information processing to work at terminals in their own homes. The vision of what has become known as the 'electronic cottage', features in all scenarios of the future of work. Although the number of people involved in this new form of homework is still small, its potential is quite large. And, according to many writers from a wide variety of political persuasions, it is a paradigmatic case for the future organization of work. According to post-industrial theory, the home as a workplace liberates people from the discipline and alienation of industrial production. Homework offers the freedom of self-regulated work and a reintegration of work and personal life. Moreover, an expansion of homework will allegedly lead to much more sharing of paid and unpaid domestic labour, as men and women spend more time at home together.

Whereas the post-industrial theorists see electronic homework as part of a positive future, for others it evokes the ugly spectre of 'sweated' self-exploitative piecework. These writers approach telework with a set of assumptions derived from the study of traditional homeworkers. They expect it to become more widespread because it is a method of production favoured by employers seeking to resist competition and protect profits by reducing wage costs. As such it is seen as part of a more general trend towards the casualization of the labour force and the growth of the informal sector. Both perspectives share a largely technologically determinist prophecy of the 'collapse of work'.

White-collar and service sector homework, both traditional and modern, have been increasing over the past two decades, even before the new information technology exerted its full influence on work arrangements. From research carried out in Europe, America and Australia, it is clear that important differences are already emerging between professional and clerical teleworkers.[7] Men predominate among the professionals, such as managerial staff, computer programmers and systems analysts, while women are the great majority of clerical workers.

Most of these are married women with young children, for whom homework is especially attractive because of their household responsibilities and the lack of affordable quality childcare. However, in practice, balancing childcare with paid work has proved difficult for many of the women as they have only limited control over a fluctuating workload. They are often employed precisely for the flexibility that this provides for employers. Like traditional homeworkers, electronic homeworkers are typically paid at piece rates and earn substantially less than comparably skilled employees working in offices, as well as having to meet their own overhead costs. Moreover, as employers do not give homeworkers employee status, they are not entitled to benefits such as sickness pay, and have no security of employment. Electronic homework for clerical women, then, is an extension of traditional homework with all its disadvantages.

The pattern of work of male professionals is quite different from that of clerical workers in that they work *from* home rather than *at* home. American research has focused on managerial and professional employees where firms turn to homework in order to retain highly qualified workers such as computer programmers. Our Australian study looked at self-employed programmers who were also able to exploit the skill shortage in their area. Most of these male professionals were earning more working at home, and many pointed to the lower overheads of running a business from home.

In our study we found that what they appreciated was not the opportunity to combine paid work with childcare but their flexible and varied working patterns. In fact, the very long hours they worked militated against any significant change to the balance between work and leisure, or work and family life. When we asked the programmers and word processor operators in our sample how working from home had changed their attitude to work, we found strong evidence of reinforced rather than transformed gender differences. Whereas the majority of men had become more work-centred, the women were more likely to have become less work-centred and more family-centred.

Thus even research on new technology homework fails to reveal simple trends. Electronic homework may well mean very different things for professional and clerical workers, and for men and women. For women clerical workers, new technology homework still reflects their labour market vulnerability - vulnerability that stems from the availability of their skill and the domestic division of labour. It is only for male professionals who possess skills which are in short supply that new technology homework presents an unambiguously attractive choice. But this hardly warrants the general enthusiasm for 'electronic cottages' that characterizes so much of the literature about the future of work.

Overall, then, new forms of computer-based homework would appear to reinforce sexual divisions in relation to paid work and unpaid domestic work, as well as to the technical division of labour. Once more we see women failing to gain the genuinely technical jobs, in this case producing software for computers. It is a stark example of the reproduction of women's traditional position in the new electronic age.

The Social Shaping of Workplace Technology

In this chapter I have been examining the impact of technological change on sexual divisions in the labour market and occupational segregation between women and men. Although new technologies may be important levers of change in the social relations of production, the gendered character of work has inhibited major transformations in the sexual division of labour. In a period of vast technological changes which have profoundly restructured work in every sphere, the resilience of the gendered character of the technical division and hierarchy of labour has been notable.

I will now turn the focus round and consider the social factors that cause technological change. The extent to which the invention and diffusion of particular technologies are themselves shaped by social forces will be explored. I will argue that the sex of the workforce and gender relations in the workplace themselves profoundly affect the direction and pace of technological change. It is only through an analysis of the processes by which technology is itself gendered that its inability to undermine gender divisions can be understood.

New technology typically emerges from modifications to and combinations of existing technology. However this is not the only force shaping technology. Industrial innovation is a product of an historically specific activity carried out in the interests of particular social groups and against the interests of others.

Technological systems are oriented to a goal and that goal is normally to reduce costs and increase revenues. When technologists focus inventive effort on the 'inefficient' components of a system, for many practical purposes inefficient means uneconomical. So technological reasoning and economic reasoning are often inseparable.

A vital issue in technical change is the cost of labour, because much innovation is sponsored and justified on the ground that it saves labour costs. In a capitalist society, class relations are a major factor affecting the price of labour. Placing the class dimension at the centre of its analysis, labour process theory is an important and well established approach to the study of technological change. Although limited with respect to gender, it provides a useful starting point for the development of a gender perspective.

Industrial Conflict and Technical Innovation

The mechanization of craft work has commonly been presented as the model for understanding major changes in the capitalist labour process. Historically, production was very dependent on the skills and knowledge of craft workers, but over the first quarter of the twentieth century their jobs were subdivided, allowing employers to dispense with skilled labour. Rather than seeing deskilling as an inexorable tendency, recent studies have emphasized the extent to which worker resistance mediated the deskilling process.[8] Craft skills provided the basis for maintaining control over the utilization of machinery and hence the basis for worker organization. A key part of this strategy was the exclusion of other non-craft workers who offered a threat to their position. As we have already seen, this mechanism of social exclusion was often deployed at the expense of women workers.

Technological innovations have played a major role in these battles for control over production.[9] In the early phases of capitalist development, machinery was used by the owners and managers of capital as an important weapon in the battle for control over production. Marx's classic account of the development of the automatic spinning 'mule' (so-called because it was a hybrid of the spinning-jenny and water-frame) in nineteenth-century Britain has, for example, been re-examined from this perspective. In the early production process of spinning the skilled adult male spinner had a central role. The spinner's centrality derived not only from his technical skills but also from his supervisory role through the system of sub-contracting labour. The spinners were highly unionized and their frequent strikes were a direct challenge to the power and profits of the cotton-masters. The self-acting mule was the employers' response to this threat.

A major strike in 1824 seems to have galvanized a number of manufacturers into recognizing their common interest in relation to the spinners. They therefore approached Richard Roberts, a well-known mechanical engineer and toolmaker. Roberts told the House of Lords Select Committee in 1851: 'The self-acting mule was made in consequence of a turn-out of the spinners at Hyde, which had lasted three months, when a deputation of masters waited upon me, and requested me to turn my attention to spinning, with the view of making the mule self-acting' (Bruland, 1982, p. 103).

The explicit purpose of this invention and its introduction was to break the power of the spinners. By changing the technology of spinning they intended to replace men on the mules with the cheaper labour of women and children. The self-actor was partially successful in its aim of curbing the spinners' militancy. In the period following the innovation, their wages were relatively depressed and strikes declined markedly. This episode exemplifies the way in which particular arenas of industrial conflict may result in the development of particular kinds of technical innovations.

In fact, the diffusion of the self-actor was relatively slow and did not have the anticipated effect of destroying the craft position of the adult male spinners. Despite radical changes in the manual component of mule spinning, these workers retained their position. The spinner-piecer system was merely replaced by an analogous minder-piecer system, which still left minders with responsibility for recruiting assistants and controlling them on the shop floor. This hierarchical division within the workforce persisted because it was the basis of the existing managerial structure in cotton spinning.

William Lazonick (1979) has shown that this reliance of the employers on a very effective form of labour management was more important than the skills or organized strength of the male minders. Thus it was the hierarchical division within the working class which conditioned technical change.

> It made it rational for capitalists to work with slightly less automated mules than were technically possible, so that failures of attention by operatives led not to 'snarls' that could be hidden in the middle of spun 'cops', but to the obvious disaster of 'sawney', where the several hundred threads being spun all broke simultaneously, with consequent loss of piecework earnings for the minder. (MacKenzie, 1984, p. 497).

The history of the self-acting mule demonstrates that an understanding of technical change as something that is based on relations of production must include an account of divisions within the working class. It not only shows how workers' resistance depends on their ability to control and restrict entry into their trade but also how employers can exploit these divisions. So the skilled worker typically looks not just in one direction – towards the capitalist who is trying to undermine his position by incorporating his skills into the machine – but also towards the mass of the 'unskilled', who can equally be seen as a threat. Typically, this will involve older, male, white workers looking in the direction of those who have at least one of the characteristics of being young, female, black or from an ethnic minority.

The development of technology cannot however simply be understood in terms of the needs of undifferentiated capital trying to control labour as an undifferentiated mass. Recent labour process work has repeatedly pointed to the weakness of assuming any simple and ubiquitous trend in the social construction of technology for control through deskilling.[10] Further it has highlighted the need to recognize differences of interest and action amongst capitalists.

The focus has shifted to the interplay between competing managerial strategies and priorities on the one hand, and various patterns of worker response on the other. There are now many documented instances where occupational competition was settled in favour of enlarged control by craft workers as well as cases where detailed control and deskilling by technology was the result

(Wilkinson, 1983). Short-term competitive pressures between capitals or motivational and flexibility concerns clearly lead to compromises over the deskilling potential of technologies.

Studies of how class relations shape technology are overwhelmingly preoccupied with traditional male unionized sections of manufacturing industry. Discussion about the impact of new technologies on the workplace has focused to a remarkable degree on the automation of machine tools. Perhaps this is because many of these male authors, like Braverman, are immersed in the romance of the skilled craftsman tragically becoming obsolete. Skilled machinists, however, have never been typical of workers and certainly women workers do not figure in their number. As there is little empirical analysis of technological development which explicitly challenges technological determinism, it is worth considering this example of a twentieth-century technology to see if lessons can be drawn from it for a gender analysis.

The Automation of Machine Tools: A Case Study of Choice

The evolution of automatically controlled machine tools is the subject of a detailed study of the design, development and diffusion of a *particular* technology, 'from the point of conception in the minds of inventors to the point of production on the shop floor' (Noble, 1984, p. xiv). This is a particularly daunting task to undertake for a modern technology when the 'heroic inventor' has left the stage to be replaced by major institutions.

The central argument of David Noble's classic study, *Forces of Production*, is that patterns of power and cultural values shape the actual processes of technological development. Noble argues that the concepts of 'economic viability' and 'technical viability', which are often used to explain technological change, are inherently political. By way of a detailed reconstruction of a lost alternative to numerical control, and by examination of variant forms of numerical control that have also vanished, Noble shows that automation did not have to proceed in the way it did. Rather, the form of automation was the result of deliberate selection.

A major goal of machine tool automation was to secure managerial control, by shifting control from the shop floor to the centralized office. There were at least two possible solutions to the problem of automating machine tools. Machining was in fact automated using the technique of numerical control. But there was also a technique of automation called 'record-playback' which was as promising as numerical control yet it enjoyed only a brief existence. Why, asks

Noble, was numerical control developed and record-playback dropped? It was the post-war period of labour militancy that provided the social context in which the technology of machine tool automation was developed.

Record-playback was a system that would have extended the machinists' skill. Although the machines were more automated under this system, the machinists still had control of the feeds, speeds, number of cuts and output of metal; in other words, they controlled the machine and thereby retained shop floor control over production.

Numerical control on the other hand offered a means of dispensing with these well-organized skilled machinists. The planning and conceptual functions were now carried out in an office because the machines operated according to computer programs. The machinist became a button pusher. Numerical control was therefore a management system, as well as a technology for cutting metals. It led to organizational changes in the factory which increased managerial control over production because the technology was chosen, in part, for just that purpose.

It would be wrong to assume that managers' goals in preferring numerical control to record–playback were necessarily realized. The introduction of numerical control on to the shop floor did not simply shift control to management. It was met with fierce resistance from the workforce. At the same time, management found that it needed to retain skilled machinists to operate the new machines effectively. Consequently, management was never able to gain complete control over production. In reality, machines do not run themselves and therefore the tendency to deskilling is always contradictory. Indeed, as Noble himself acknowledges, the subsequent development of machine tool technology has made it technically feasible and potentially economical to institute shop floor programming. As technological advances opened up new areas of application - in smaller firms involved in small batch or specialised production - it also provided opportunities for craft workers to regain control over programming. In fact, the operational requirements of these small firms may be more compatible with shop floor programming than with a managerial strategy oriented around deskilling.

Noble's study is remarkable for its attempt to encompass many different levels of social determination of the technology. It does not simply rely on treating technology as being determined by management's demand for control over workers. It goes beyond that to include the role and interests of the military in that post-war period, as well as the ideology and interests of engineers. Although it was the social relations of production that tipped the balance in the choice of technology in Noble's example, he demonstrates how the demand for management control coincided with the command and control goals of the military. He also shows how the ideology and interests of engineers who take the view that the most automated is the most advanced and that the human element should be eliminated from production because it is the potential source of 'human error' fits in with the idea of management control.

However, while emphasizing the various class forces that shape the design and application of machine tools, Noble fails to consider that there is also a gender dimension to these forces. This could have been observed through the role of the state, capital and unions but it is particularly evident in his otherwise excellent account of the ideology and culture of engineers. Engineering culture, with its fascination with computers and the most automated techniques, is archetypically masculine and would have provided an excellent opportunity for an integration of class and gender perspectives on technological change.

The Gendered Relations of Technology

Class divisions have been central to the analysis presented so far, but the relations between women workers and men workers are of fundamental importance for any discussion of the development of technology. One of the ways that gender divisions interact with technological change is through the price of labour, in that women's wage labour generally costs considerably less than men's. This may affect technological change in at least two ways. Firstly, as we have seen, employers may seek forms of technological change that enable them to replace expensive skilled male workers with low paid, less unionized female workers. Secondly, because a new machine has to pay for itself in labour costs saved, technological change may be slower in industries where there is an abundant supply of cheap women's labour.

There is some historical evidence that the rate of technical development has depended, at least in part, on the price and skill flexibility of the available labour force. For example, the clothing industry has remained technologically static since the nineteenth century with little change in the sewing process. There are no doubt purely technical obstacles to the mechanization of clothing production, such as the floppy material involved and changing styles and fashions. However, leaving aside the technical difficulties, there will be less incentive to invest in automation if skilled and cheap labour power is available to do the job.

Thus there is an important link between women's status as unskilled and low paid workers, and the uneven pace of technological development. Traditionally it is women who sew and they have been available for low wages, either in Third World countries or as migrant labour in Western capitalist countries. The fact that clothing workers are regarded as unskilled is due in large measure to their lack of industrial strength, which is in turn due to the large pool of women whose social situation forces them to compete in this area of work. It is not possible for anybody to sit down at a sewing machine and sew a garment without previous experience. To be a competent machinist demands considerable knowledge and experience with the machine. Although this is one area where women are at ease with machines, this is seen as women's supposed natural aptitude for sewing and thus this technical skill is devalued and underpaid.[11]

There is a more direct sense, however, in which gender relations leave their imprint on technology. As I pointed out earlier, recent feminist work has emphasized that distinctions of skill between women's and men's work have as much to do with job control and wage levels as they have to do with actual technique. However, this formulation understates the tangible basis of skill. Men selectively design tools and machinery to match the technical skills they have cultivated. Machinery is designed by men with men in mind; industrial technology reflects male power as well as capitalist domination.

The Transformation of Typesetting: Building in Sex-Bias

The best examples of the gendering of technology come from Cockburn's (1983) history of typesetting, which provides a detailed description of the technological evolution of the computerized photocomposition system. Like Noble, she shows that automation did not have to proceed in the way that it did. Rather, the form of automation was the result of 'deliberate' selection. Cockburn suggests that the technical choices made can only be understood by looking closely at the conflictual relations of production, including the central role of gender relations.

Computerized photocomposition technology has what is known as a QWERTY keyboard. Q–W–E–R–T–Y are the characters on the second top row left-hand side of a conventional typewriter. This is now the standard keyboard incorporated into computers. However there was nothing inevitable about this. Electronic circuitry is in fact perfectly capable of producing a Linotype lay on the new-style board. Linotype Paul have manufactured one. The lay of the Linotype keyboard differs greatly from QWERTY. Not only does it have 90 keys in contrast to 44, the relative position of the letters of the alphabet also differs from that of a typewriter and the keys are larger and spread further apart.

So what politics lie behind the design and selection of this keyboard? In choosing to dispense with the Linotype layout, management were choosing a system that would undermine the skill and power basis of the Linotype operators, the highest paid of all the craftsmen. All the operators would be reduced to novices on the new board, as the inputting would now require little more than good typing ability. This would render typists (mainly women) and Linotype operators (men) equal competitors for the new machines; indeed, it would advantage the women typists. The QWERTY technique was designed with an eye to using the relatively cheap and abundant labour of female typists.

The history of mechanized typesetting offers another instance of clear sex-bias within the design of equipment. A nineteenth-century rival to the Linotype was the Hattersley typesetter. Compositors hated technical systems such as the Hattersley typesetter that separated the jobs of composing and distribution. It had a separate mechanism for distributing type, designed for use by girls. The separation of the setting (skilled) and distribution (the unskilled job of putting the letters back in their letter box) was devised as a means of reducing overall labour costs. Compositors feared that employers would try to expand this use of cheaper, unskilled labour once it got a foothold into the composing room.

The Linotype machine on the other hand did not represent the destruction but merely the mechanization of the compositors' setting skills as a whole. The key aspect of this successful machine was that it eliminated distribution as a task - since letters were formed anew each time by the action of brass mould on molten metal. After the type was used it was simply melted down ready to be re-used. The compositors actually welcomed the Linotype machine because it did not depend for its success on the employment of child labour. The men's union, the London Society of Compositors, even wrote a letter to the Linotype Company Ltd. in 1893 congratulating them: 'The Linotype answers to one of the essential conditions of trade unionism, in that it does not depend for its success on the employment of boy or girl labour'. On the contrary, by cutting out the task of distribution, it stopped any possible inroads that boys and women might make into the trade. Thus, in deference to the organizational strength of the union, the Linotype manufacturing company adopted a technology that was beneficial to the union men.

Perhaps, finally, there is another level on which the technology of production reflects male power. Feminists have understandably tended to under-emphasize the material realities of physical power, given that women's exclusion from numerous occupations has been legitimated in biological terms. It is often still said that men are naturally stronger and therefore more suited to certain types of work. However, as Cockburn (1983, p. 203) correctly stresses, 'the construction of men as strong and capable, manually able and technologically endowed, and women as physically and technically incompetent' is a social process. It is the result of different childhood exposure to technology, the prevalence of different role models, different forms of schooling, and the extreme sex segregation of the job market. The effect of this is an implicit bias in the design of machinery and job content towards male strength.

In composing work, the lifting and carrying of the forme is a case in point. The forme is heavy and in fact beyond the strength of not only women but also many men, particularly older men and younger apprentices. However, by defining this task as one that requires muscle, women workers cannot threaten to undercut men's labour. The size and weight of the forme is in fact arbitrary. Printing presses and the printed sheet could have been smaller too. Tradition alone has decided at what weight the use of hoists and trolleys to transport the forme is introduced. There is nothing natural about units of work. Whether it is hay bales or 50-kilo bags of cement or plaster, they are political in their design. Capitalists and workers have a political interest in the design of work processes. Employers prefer workers to use their brawn when it leads to more efficiency and lower production costs. Male workers use their bodily and technical effectivity to design machinery and work tasks so as to constitute themselves as the capable workers and women as inadequate.

It is overwhelmingly males who design technological process and industrial machinery. It is the knowledge and experience of engineers and of the workers who use the machines which filters through into the shape of new technologies. Mechanical equipment is often manufactured and assembled in ways that make it just too big and heavy for the 'average' woman to use. This need not be a conscious process or conspiracy. It is rather the outcome of a pre-existing pattern of power. This is not to imply that men always design technology for their own use and in their own interests. It is more complex than that. Capital's interest cannot be supposed always to coincide with that of men as a sex. As we have seen, some technologies are designed for use by women in order to break the craft control of men. Thus gender divisions are commonly exploited in the power struggles between capital and labour. In this way, the social relations that shape technology include those of gender as well as class.

This chapter has argued that despite technology being seen as a driving force it has not ushered in a new order but rather has been built into the pre-existing relations of sex, class and race that structure the labour force and employment opportunities. Although there has been an expansion of job opportunities for women in some of the new information services, 'women's jobs' and 'men's jobs' are as strongly demarcated as ever. This is in part because social relations are expressed in and shape technologies. The pace and direction of technological development reflect existing gender relations as much as they affect the sexual division of labour.

NOTES

1 For a critical review of this literature, see Pollert (1988) who argues that what she calls the 'left-reformist' writing which advocates 'flexible specialization' as a panacea has a conceptual affinity with management literature on the 'flexible firm'.
2 Liff (1988) and Webster (1989) argue along similar lines in their excellent reviews of the literature on office automation. See also the major two-volume report commissioned by the American National Research Council, edited by Hartmann et al. (1986, 1987), which examines the effects of technological change on both the quantity and quality of women's employment, particularly in clerical work.
3 I do not want to rehearse the various arguments contained in this literature, as several comprehensive reviews already exist. See Thompson (1983) for an excellent introduction to debates on the labour process.
4 To anticipate my argument about the gendered design of technology, Juliet Webster has pointed out to me that the first word processing systems were deliberately restricted to stand-alone, dedicated micro-computers rather than being developed in package form to be used on mainframes or all-purpose micros. This particualr design was aimed at the main users, women office workers who had formerly worked with typewriters. Word processing machines were given dedicated keyboards with text editing functions embedded in the hardware, and screens which imitated pieces of paper, to make them resemble their mechanical

predecessors. These features made them seem more accessible to women with skills and experience in operating office equipment (pink technology) as opposed to computers (blue technology).

5 See Crompton and Jones (1984).

6 Game and Pringle (1983) and Cockburn (1985) both present case studies of employment where technical developments have substantially altered the skill and task range of jobs and yet the distinctions between 'men's' work and 'women's' remain, with men still monopolizing the technical jobs. See Purcell (1988) for a review of recent British research on the gendering of occupations.

7 See Wajcman and Probert (1988) for a report of our Australian study on new technology homework, which includes a general literature review.

8 See Thompson (1983, chapter 4) for an extensive discussion of the deskilling debate.

9 This argument is developed at greater length in MacKenzie and Wajcman (1985); especially see the extracts by Bruland and Lazonick.

10 See Elger's (1987) review of several recent studies.

11 For a useful discussion of the undervaluing of skills in the clothing industry, see chapter 5 of O'Donnell (1984).

3 Reproductive Technology: Delivered into Men's Hands

Nowhere is the relationship between gender and technology more vigorously contested than in the sphere of human biological reproduction. Women are the bearers, and in most societies the primary nurturers, of children. This means that reproductive technologies are of particular significance to them. Birth control has been a major issue for all movements for women's equality, and much feminist scholarship has been devoted to uncovering women's struggle throughout history against the appropriation of medical knowledge and practice by men.

Central to this analysis and of increasing relevance today is the perception that the processes of pregnancy and childbirth are directed and controlled by ever more sophisticated and intrusive technologies. Implicit in this view is a concept of reproduction as a natural process, inherent in women alone, and a theory of technology as patriarchal, enabling the male domination of women and nature.

The burgeoning debate about these issues has largely been conducted within the feminist movement on the one hand and within the fundamentalist Right on the other. Interestingly, socialists have generally been silent on recent developments in reproductive technology, perhaps because they primarily affect women, or perhaps because they do not concern workplace production, the Left's traditional obsession. But these are the technologies of life, raising complex moral issues about the role of human intervention in the world of living beings. This chapter will explore feminist perspectives on reproductive technologies, placing them in the wider context of the growing supremacy of technology in medicine.

Feminist Perspectives on Reproductive Technology

The literature on reproductive technology is rife with technological determinist arguments which assume that changes in technology are

3.1 *Source:* Recycled Images

the most important cause of changes in society.[1] Perhaps here more than elsewhere, major technological advances are seen as having directly transformed women's lives for the better. The technologies of pregnancy and childbirth are said to have put an end to the dangerous and painful aspects of giving birth. Healthy pregnancies and healthy babies are attributed to the wonders of modern antenatal care, now a highly medicalized and technologized process. The new sophisticated techniques for monitoring foetal development in the early stages of pregnancy mean that some 'defective' foetuses can be aborted. Infertile women who previously had no options can now embark on infertility programmes that promise the chance of conceiving 'naturally'. And, most common of all, advances in the technologies of fertility control are seen as the key to the massive social changes that have occurred for women's equality. The widespread availability of reliable contraception and abortion, a right often fought for by women, have meant that for the first time in human history women are in control of their own bodies.[2]

Technology as the Key to Women's Liberation

In the early period of the contemporary women's movement, reproductive technology was seen as particularly progressive because it opened up the potential for finally severing the link between sexuality and reproduction. The much-cited advocate of the use of high technology to liberate women was Shulamith Firestone. In *The Dialectic of Sex* (1970) she emphasized the need to develop effective contraceptive and birth technologies in order to free women from the 'tyranny of reproduction' which dictated the nature of women's oppression. Patriarchy was seen to be fundamentally about the control of women's bodies, especially their sexuality and fertility, by men. This view located women's oppression in their own biology and posited a technological fix in the shape of ectogenesis. The application of a neutral technology would bring an end to biological motherhood and thus make sexual equality possible.

Since then, feminist analysis has not shared Firestone's enthusiasm for the artificial womb as the key to women's liberation. Instead, feminists have recently been more concerned either to oppose the experimentation on women's bodies that the development of these techniques entails or to harness these techniques in the interests of fulfilling women's maternal desires.

Genetic research, bio-technology and infertility treatment are now making such dramatic advances that Firestone's ideas no longer seem to belong in the realm of fantasy. The organic unity of foetus and mother can no longer be assumed now that human eggs and embryos can be moved from body to body or out of and back into the same female body. The major proponents of the possibilities of reproductive technologies are the scientists and medical practitioners developing the techniques as well as women who have benefited from them. Leading infertility doctors argue that embryo research promises the possibility of eliminating some of the most crippling forms of hereditary disease and most importantly, gives hope to previously childless couples. As one Member of Parliament recently put it:

> The object of our interest in medical research into embryology and human fertilisation is to help humanity. It is to help those who are infertile and to help control infertility. . . . The researchers are not

> monsters, but scientists. They are medical scientists working in response to a great human need. We should be proud of them. The infertile parents who have been helped are grateful to them. (Pfeffer, 1987, p. 81)

However, all over the world, the use of human embryos in scientific research is becoming a major source of controversy. Governments are under pressure to impose tighter regulation and define the limits of what is permissible. Ethical and religious objections have been strongly voiced to the inexorable advance of science and technology into the sacred realms of creation. The 'right to life' lobby calls for legislation to ban research in human embryology and the practice of in-vitro fertilization. Just as they oppose abortion as an unnatural interference with procreation, their concern is for the life and soul of the foetus. The intense public debate is centred around the question of which, if any, of the procedures and experimental programmes should be licensed and given resources.

In Australia, Europe and North America there is growing debate among feminists over the impact that these novel reproductive and genetic technologies will have on women's lives. This is a very divisive area for feminists. Whereas abortion and contraception were about challenging the traditional definition of femininity which equated it with motherhood, by contrast these new technologies are about fulfilling, rather than rejecting, the traditional feminine role.

A shared concern is that techniques such as in-vitro fertilization coexist with a powerful ideology of motherhood. Many feminists argue that the in-vitro fertilization programme reinforces the definition of motherhood as a biological imperative rather than a social relationship. As Christine Crowe (1987, p. 84) observes: 'IVF does *not* cure infertility; it provides (and for a few women only) an avenue to biological motherhood through technological intervention'.[3] It is a 'technological fix' in the sense that it does not at any point address the initial causes of infertility. Doctors and the media describe these technologies as enhancing women's 'natural need' to mother, and infertile women as desperate. Much of the feminist discussion centres on the notion of choice and whether the right to choose to have an abortion can be equated with the right to choose to have a child.[4] As we shall see below, feminist support for techniques such as in-vitro fertilization is founded in the belief that these technologies increase women's choices and that women do indeed have the right to reproduce.

Reproductive Technology as Patriarchal Domination

Most vocal in their opposition to the development and application of genetic and reproductive engineering are a group of radical feminists who in 1984 formed FINRRAGE (Feminist International Network of Resistance to Reproductive and Genetic Engineering). Represented by authors such as Gena Corea (1985), Jalna Hanmer (1985), Renate D. Klein (1985), Maria Mies (1987) and Robyn Rowland (1985), they see the development of reproductive technologies as a form of patriarchal exploitation of women's bodies.

Whereas Firestone saw women's reproductive role as the source of their oppression, FINRRAGE writers want to reclaim the experience of motherhood as the foundation of women's identity. For, as Robyn Rowland (1985, p. 78) expresses it: 'the qualities of mothering or maternal thinking stand in opposition to the destructive, violent and self-aggrandizing characteristics of men.' The previously celebrated technological potential for the complete separation of reproduction from sexuality is now seen as an attack on women. Radical feminist theory sees these techniques as an attempt to appropriate the reproductive capacities which have been, in the past, women's unique source of power. It is about removing 'the last woman-centred process from us'. For Jalna Hanmer (1985, p. 103), 'The dominant mode of [patriarchal] control is changing hands from the individual male through marriage to men as a social category, through science and technology. . . . The locus of control and struggle is shifting from sexuality to reproduction and childcare, i.e. motherhood.'

For this group of feminists, who have criticized the ways in which patriarchal society has ignored or sanctioned sexual and domestic violence against women, the new reproductive and gene technologies are 'violence against women in yet another form'. 'Genetic and reproductive engineering is another attempt to end self-determination over our bodies'. According to this theory, techniques such as in-vitro fertilization, egg donation, sex predetermination and embryo evaluation offer a powerful means of social control because they will become standard practice. Just as other obstetric procedures were first introduced for 'high risk' cases and are now used routinely on most birthing women, these authors fear that the new techniques will eventually be used on a large proportion of the female population.

FINRRAGE sees reproductive technologies as inextricably linked with genetic engineering and eugenics. It is techniques such as in-vitro fertilization which provide researchers with the embryos on which to do scientific research. A parallel is drawn between the way in which men have been increasingly controlling the reproduction of animals to improve their stock by experimenting on them, and the extension of this form of experimentation to women. The female body is being expropriated, fragmented and dissected as raw material, or providing 'living laboratories' as Renate Klein puts it, for the technological production of human beings.

The most powerful statement of this is Gena Corea's image of 'the reproductive brothel' which extrapolates from the way animals are now used like machines to breed, to a future in which women will become professional breeders, 'the mother machine' at men's command. Some writers argue that these techniques will actually replace natural reproduction, guaranteeing the fabrication of genetically-perfect babies. According to this futuristic dystopia, men will achieve ultimate control of human creation and women will be redundant.

Many feminists have explained the patriarchal desire for control over reproduction in psychoanalytic or psychological terms, associating it with male fear of female procreativity and the quest for immortality. The potential of this technology to disconnect the foetus from a woman's body is seen as a specific form of the ancient masculine impulse 'to confine and limit and curb the creativity and potentially polluting power of female procreation' (Oakley, 1976, p. 57), in short, male womb envy. Embedded in this approach, and most explicit in the work of Maria Mies, is a conception of science and technology as intrinsically patriarchal. FINRRAGE states that they want a new feminist science based on 'a non-exploitative relationship between nature and ourselves'. Clearly, feminist philosophical theorizing about the masculinist character of scientific objectivity and rationality is being heavily drawn on in current debates about reproductive technologies.

Mies argues that it makes absolutely no difference whether it is women or men who apply and control this technology; this technology is intrinsically an instrument of domination, 'a new stage in the patriarchal war against women'. Technology is not neutral but is always based on 'exploitation of and domination over nature, exploitation and subjection of women, exploitation and oppression of other peoples' (1987, p. 37). Mies argues that this is the very logic of the natural sciences and its model is the machine. For her the method of technical progress is the violent destruction of natural links between living organisms, the dissection and analysis of these organisms down to their smallest elements, in order to reassemble them, according to the plans of the male engineers, as machines. The goal of the enterprise is to become independent of the 'moods' of nature and of the women out of whom life still comes. Reproductive and genetic technologies are about conquering the 'last frontier' of men's domination over nature.

Reproductive Technology as Neutral

Rather than seeing reproductive technologies as a sustained attack on women, another group of feminist commentators emphasize the ambivalent effects that reproductive technologies have on the lives of women. According to Michelle Stanworth (1987, p. 3), a blanket rejection of these innovations is inadequate as many of them 'offer indispensable resources upon which women seek to draw according to their circumstances'. These new technologies are seen as having the potential to empower, as well as to disempower, women and the discussion is couched in terms of 'the costs and the benefits'.

These authors argue that the women's movement has largely ignored the problems of infertility and treated women who participate in these high-tech research programmes as 'blinded by science' and as passive victims of pronatal conditioning. According to them, most of the authors associated with FINRRAGE fail to consider women as active agents who have generated demands for such technologies because of their authentic desire to bear children. As a result, feminist opposition to these technologies has a tendency to 'confuse masculine rhetoric and fantasies with actual power relations, thereby submerging women's own response to reproductive situations in the dominant (and victimizing) masculine text' (Petchesky, 1987, p. 71). Reproductive technologies may be the only opportunity infertile women have to fulfil this need and therefore we should support their 'right of reproductive choice'.

This group of writers take issue with the radical feminist view that technologies in themselves have patriarchal political properties. Instead, they problematize the institutional setting in which these medical/technical procedures occur. Whereas the FINRRAGE authors are against these innovations because they inevitably disempower women, according to Rosalind Pollack Petchesky, 'we need to separate the power relations within which reproductive technologies . . . are applied from the technologies themselves' (1987, p. 79). Similarly, for Michelle Stanworth, the problem is not technology but the way 'these technologies draw their meaning from the cultural and political climate in which they are embedded' (1987, p. 26).

The feminist debate about the new reproductive technologies

reviewed here is a relatively recent one and, as a result, it is characterized by more sensitivity to 'the politics of difference' than some of the earlier feminist literature. There is now a much clearer realization that gender, that what it is to be a woman, is experienced everywhere through such mediations as sexual orientation, age, race, class, history, and colonialism. The recognition that new technologies may have very different implications for Third World and First World women, within and between countries, is a strength of much of the literature.

The real dangers for women that accompany medical and scientific advances in the sphere of reproduction are directly related to the different circumstances of women's position in society. Access to the benefits of expensive techniques such as in-vitro fertilization is heavily related to the ability to pay. Women who are poor and vulnerable will not have access to these techniques and furthermore, they will be least able to resist abuses of medical power and techniques. For example, ethical issues over sex predetermination have a special urgency given evidence that the technique of amniocentesis is currently being used to preselect female fetuses for abortion in India. Sterilization and drugs such as Depo-Provera, as well as hazardous experiments, have been particularly targeted at coloured women.

The potential use of increasingly sophisticated forms of genetic screening is likely to influence the definition of a 'genetic defect' and may have implications for the way disability is seen in society. Research such as Wendy Farrant's (1985) shows that the medical management of prenatal screening in Britain has taken the form of gaining women's consent for termination as a condition for being allowed the amniocentesis test. In this context, these techniques are about population control rather than about enabling women to make more informed choices about reproduction.

There is broad agreement among feminists about these dangers. For those feminists who dispute the FINRRAGE analysis, these dangers are seen not as a function of the technologies themselves, but of their abuse. This position is summed up by Stanworth (1987, p. 15), when she says that these technologies have been 'a double-edged sword. On the one hand, they have offered women a greater technical possibility to decide if, when and under what conditions to have children; on the other, the domination of so much reproductive technology by the medical profession and the state has enabled others to have an even greater capacity to exert control over women's lives'. From this perspective, therefore, the feminist critique of reproductive technologies goes no further than demanding access to knowledge and resources so that women are able 'to shape the experience of reproduction according to their own definitions'.

An aspect of the politics of reproductive technology left out by this account is that the technologies redefine what counts as illness. 'Infertility' now becomes not a biological state to which the woman must adapt her life, but a medical condition - a problem capable of technological intervention. The very existence of the technologies changes the situation even if the woman does not use them. Her 'infertility' is now treatable, and she must in a sense actively decide not to be treated. In this way the technologies strengthen the maternal function of all women, and reinforce the internalization of that role for each woman.

Indeed, the emphasis placed on women's right to use these technologies to their own ends tends to obscure the way in which historical and social relations are built into the technologies themselves. While recognizing the social shaping of women's choices in the sense of motivations, few participants in the debate see that the technologies from which women choose are themselves shaped socially.[5]

Techniques such as in-vitro fertilization, egg donation, artificial insemination, and surrogacy have the potential to place the whole notion of genetic parenthood, and thus family relationships, in jeopardy. However, only those technologies that reinforce the value of having one's 'own' child, one that is genetically related to oneself, are being developed and, as Patricia Spallone (1987, pp. 173–4) argues, these values determined the Warnock Committee's assessment of 'acceptable' risks to women's health. Despite the dangers, the Committee approved the use of in-vitro fertilization, where egg donation provides an offspring which is genetically related to the husband. Yet the technique of egg donation by uterine lavage (embryo flushing or surrogate embryo transfer) was rejected on the basis of physical risks. The medical risks involved in this procedure are no greater, but it carries the risk of unwanted pregnancy in the donor woman. Two women would then be sharing a pregnancy and the existence of this donor mother-to-be would challenge the usual categories of motherhood. This technology was rejected, not on the grounds that it endangers women's health, but because of its socially disruptive character to the identification of blood ties with the family.

Women are in fact selecting from the very restricted range of technological options which are available to them. This is glossed over by the feminist critics of the FINRRAGE position. By focusing on the sexual politics in which the new reproductive technologies are embedded, they pay insufficient attention to the technology itself. In adopting, implicitly or explicitly, the use/abuse model of technology, they fail to appreciate the extent to which technologies have political qualities. This is where the strength of the FINRRAGE analysis lies. In my view, FINRRAGE are right to argue that gender relations have profoundly structured the form of reproductive technologies that have become available.

To make this claim however one does not need to conceptualize it in terms of a monolithic male conspiracy. As Langdon Winner (1980, p. 125) has said: 'to recognize the political dimensions in the shapes of technology does not require that we look for conscious conspiracies or malicious intentions'. Nor does it imply that men are a homogeneous group. While it is evident that all the stages in the career of a medical technology, from its inception and development, through to consolidation as part of routine practice, are a series of interlocking male activities, the male interests involved are specifically those of white middle-class professionals. The division of labour that produces and deploys the reproductive technologies is both sexual and professional: women are the patients, while the obstetricians, gynaecologists, molecular biologists and embryologists are men.

If we regard technology as neutral but subject to abuse we will be blinded to the consequences of artefacts being designed and developed in particular ways. To make sense of reproductive technology we need to examine the social and economic forces that drive research forward or that inhibit more progressive developments. Throughout this book I have argued that certain kinds of technology are inextricably linked to particular institutionalized patterns of power and authority, and the case of reproductive technologies is no exception. Men's appropriation of technology here, as in the other areas we have examined, has been decisive in attempts to create and maintain control over women. This can best be demonstrated by looking at the emergence of specific technologies and how they figure in the historical establishment of male hegemony in Western medicine.

The Medicalization and Mechanization of Childbirth

Delivered into Men's Hands

A major focus of feminist historians of medicine has been to document the central role of women healers and midwives before the rise of modern medicine. Up until the close of the seventeenth century attendance on childbirth had always been the preserve of women, traditionally providing a livelihood for the wife and widow. It was midwives who came to women in labour and who assisted women in the process of giving birth. Their experience and knowledge about birthing and about birth assistance was passed from one generation of women to the next. Throughout the eighteenth century a bitter and well documented contest took place between female midwives and the emerging male-dominated medical profession, as to who would have control over intervention in the birth process (Ehrenreich and English, 1979 and Donnison, 1977). It emerges from these accounts that a particular technology played a crucial role in determining the outcome.

In England from the 1720s onwards an increasing number of men were entering midwifery in direct competition with women. Before that time, surgeons (an exclusively male occupation) had only been called in for difficult cases where natural delivery was not possible. They had carved out this work in the thirteenth century by forming surgeons' guilds which gave them the exclusive right to use surgical instruments. Before the invention of forceps however there was little they could do except to remove the infant piece-meal by the use of hooks and perforators, or to perform a Caesarian section on the body of the mother after her death (Donnison, 1977). Obstetric forceps were introduced by the Scottish apothecary William Smellie by the 1730s (see figure 3.2).

The forceps enabled its user to deliver live infants in cases where previously either child or mother would have died, and also to shorten tedious labour. According to custom, midwives were not allowed to use instruments as an accepted part of their practice. The use of forceps thus became the exclusive domain of physicians and surgeons, and was associated with the emerging profession of medicine. The introduction of forceps gave these men the edge over female midwives who were adept at the manual delivery of babies and who had all the practical knowledge about birth and birthing. As soon as this technology was introduced it was seized upon by physicians, who used it far too often, even in the contemporary opinion of the inventor himself. The outcome of the struggle that ensued was that the midwives lost their monopoly on birthing intervention, which became the province of the profession of medicine. For the first time in history childbirth, which had always been 'women's business', had been captured by men.

Clearly the ascendancy of male obstetrics was the result of a number of interrelated factors, a critical element being the movement of childbirth from home into the newly established lying-in hospitals. However, the invention of one of the first technological aids to birthing provided a crucial resource for male medical practitioners. It is telling that the public debate precipitated by the entry of men into

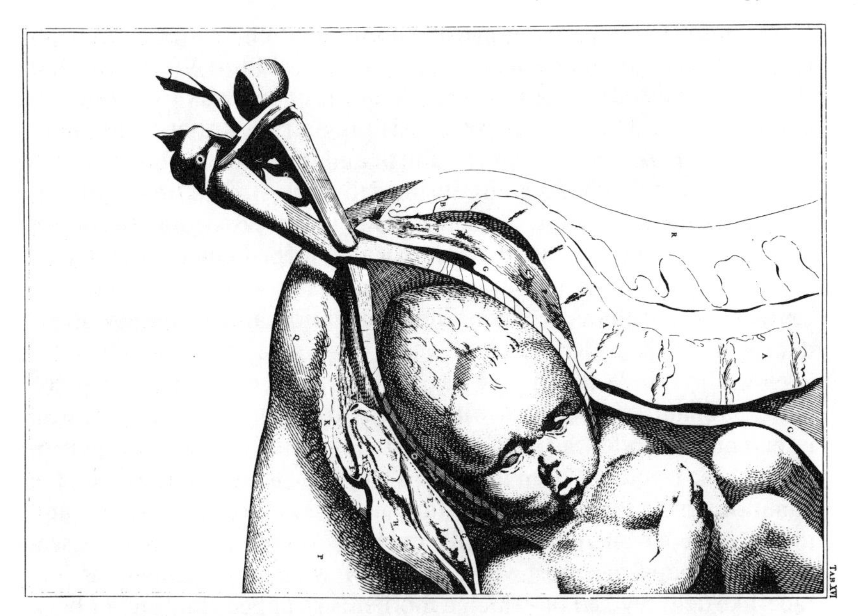

3.2 Obstetric forceps as 'artificial hands'
Source: William Smellie, *A Sett of Anatomical Tables*, 1754, plate 16.

midwifery pivoted around the use of instruments such as obstetric forceps: 'the doctors' practice of midwifery was becoming distinguishable by its very technical aspect' (Faulkner, 1985, p. 93). Young male midwives were often incompetent and frequently used instruments unnecessarily to hasten the birth and save their time, often damaging the mother and killing the child. The misuse of instruments was still common enough to attract the following criticism from a leading medical practitioner, a James Blundell of Guy's Hospital, who wrote in 1834 that some men seemed to suffer from 'a sort of instinctive impulse to put the level and the forceps into the vagina' (Donnison, 1977, p. 50). Thus technical intervention rapidly became the hallmark of male medical practice.

This is not to say that birthing women were necessarily hostile to increased technical intervention. In the early decades of this century there was considerable feminist agitation in favour of the use of anaesthesia during labour, which male physicians were then opposing. Women took up the cause of drug-induced 'twilight sleep' because they saw it as 'the newest and finest technique available' to relieve the acute pain of childbirth. Physicians' objections to its use took various forms but ultimately they were defending their professional prerogative to determine the patient's treatment. As Judith Walzer Leavitt shows, this episode is a good example of the complexity of arguments about control. The doctors were resisting a process that would have reinforced *their* control over childbirth and the women were demanding the right to be unconscious during delivery! Although the twilight-sleep movement was motivated by a desire to increase women's control over the birthing process, it paradoxically 'helped change the definition of birthing from a natural home event, as it was in the nineteenth century, to an illness requiring hospitalization and physician attendance' (1986, p. 140).

Nowadays, in Western societies, childbirth is generally experienced in hospital and is associated with increased and routine technological intervention. Under the aegis of the predominantly male medical profession, the trend has been towards the routine use of anaesthesia, the common resort to forceps, the standard practice of episiotomy, and the increase in births artificially induced as well as Caesarian sections.[6] Perhaps the most vivid image of women's treatment is 'the rack-like delivery bed on which a mother is strapped, flat on her back with her legs in stirrups, in a position which might have been deliberately designed to make her own efforts to bear a child as ineffectual as possible' (Donnison, 1977, p. 198). A number of feminist authors, including Ann Oakley, have argued that this medical 'management' of pregnancy and childbirth by a powerful professional male elite has reduced women to the status of reproductive objects, engendering adverse emotional experiences for childbearing women. Contemporary feminists have been particularly critical of the extent to which birth has been transformed from a natural process into a pathological one.

Until fairly recently it was generally assumed that maternal and neonatal deaths were reduced as a direct result of the increased proportion of hospital confinements and the application of technology in pregnancy, labour and birth. This belief explains women's apparent tolerance for a system that some have argued has transformed birth into a passive and alienating experience. It is now widely acknowledged however that in many, if not in most, cases, massive technological intervention in childbirth is unnecessary. With the exception of risky births and women who need Caesarian sections, such intervention is not a biological necessity; rather, it reflects the structure of power and decision making within obstetrical situations.

Recent sociological and medical literature has been reevaluating the contribution of medical technology to the health of mothers and babies, in comparison with social factors such as the standard of nutrition and sanitation. As Jill Rakusen and Nick Davidson (1982, p. 152) put it: 'The single most significant contribution to a cut in the death and handicap rate among newborn babies would be a comprehensive anti-poverty programme'.[7] Indeed, the women whose welfare might be most enhanced by these medical technologies have least access to them.

The strength of the feminist critique of professional medical care is not only its dissection of medical-technological treatments but its analysis of the way scientific and medical knowledge is itself gendered. To understand the medical treatment of birth, it is important to recognize that in the development of Western thought and medicine, the body came to be regarded as a machine. The Cartesian model of the body as a machine and the physician as technician or mechanic emerged in the seventeenth century and was integral to the development of the biomedical sciences. This mechanical metaphor continues to dominate modern medical practice and underlies the propensity to apply technology and to see surgery as the appropriate cure.

As I noted in chapter 1, gender symbolism and representations of sexual difference were central to the scientific and medical texts of the eighteenth and nineteenth centuries. In contrast to the male norm, women's bodies were depicted as frail and prone to physical and mental disease, the prime objects of medical intervention. Ludi Jordanova's recent book *Sexual Visions* contains fascinating material on the depiction of the differences between women and men in the biomedical sciences between the eighteenth and twentieth centuries. These sciences were associated with the idea of the unveiling of nature, and woman, as the personification of nature and 'the other', was thus the appropriate corpse for the male practice of anatomy. (Although imagery of women's bodies was predominant, the unveiling of 'otherness' also took a racial form.) Jordanova argues that gender is still a central medical metaphor and by examining advertisements in a contemporary medical magazine she explores the ways in which illnesses are visually tagged as 'male' or 'female'. 'Depression, anxiety, sleeplessness and migraine are likely to be associated with women, while disorders that can inhibit full movement and strenuous sporting activities are associated, metaphorically, with masculinity.' (1989, p. 144)

The language of the biomedical sciences today is no less suffused with implicit assumptions about and imagery of sexual difference. Through a detailed comparison of medical writings on the female reproductive system with those on the male equivalent, Emily Martin

(1987) found that the cultural grammar was radically distinct. Whereas the dominant metaphors used of the female system are negative and demeaning to women, by contrast those used of the male system suggested power and positive qualities. The images contained in medical descriptions of menstruation and menopause are characteristically in terms of failed production, breakdown and decay; sperm is depicted as 'amazing . . . in its sheer magnitude'.[8] In obstetric literature, the uterus is regarded as the machine that produces the baby and the woman is the labourer supervised and managed by the doctor. Even in the act of conception the language of medicine assigns a passive role to women and an active role to the male. Propelled by a powerful tail, sperm, that 'nuclear war-head of paternal genes', actively swims upstream to fertilize the waiting egg.

The profound gender-bias in the way medical science views women's and men's bodies, in its very way of seeing problems, has consequences both in the rate and kind of technological intervention. This is exemplified in the differential treatment of reproductive disorders in women and men. Infertility treatment is primarily aimed at women and male infertility is hardly visible or even acknowledged. 'Unlike the female reproductive system, which is served by gynaecology, there is no medical specialty for the male reproductive system' (Pfeffer, 1985, p. 35). Far from being a sign of neglect, this is symptomatic of the medical profession's refusal to see the male reproductive system as defective.

Much medical technology has no doubt been of physical benefit to women, particularly in terms of pain relief, and this has been underemphasized in the feminist literature, which is highly critical of modern, hospital-based obstetric practices. This view equates the increase in technological intervention with a corresponding loss of women's power and control over a dehumanized birth process. The history of reproductive technology is thus seen in terms of the oppression of women by science and medicine. Modern practices are compared unfavourably with explicit or implicit conceptions of what childbirth was like in earlier periods or in primitive societies. It is presumed or asserted that until the advent of male medical control, childbirth was a safe, non-alienating and purely 'natural' physiological process; that women midwives and relatives attended in a sympathetic and supportive role.

However, as Sally Macintyre (1977, p. 18) points out: 'Childbirth is, of course, socially controlled in all societies.' Far from women themselves being individually in control, childbirth is invariably surrounded by rules, customs, prescriptions and sanctions. Indeed, historically and cross-culturally it is evident that women commonly police the process themselves, not simply deferring to the expectant mother's own wishes. To counterpose masculine technologized child-birth to women's 'natural' ways begs the question. The issue is not what childbirth was or would be like for women without the controls imposed by modern technology, but why the technologies we have take the form they do. Thus we need to look at the social context in which the new reproductive technologies have developed.

Technology and Professionalization

In all professions, claiming expert technical knowledge has been favoured as a way of legitimating specialization. The unequal power relations between medical practitioners and their female patients are based on a combination of factors, predominantly those of professional qualification and gender. Oakley argues that the technological imperative within reproductive medicine is intrinsic to the defence of doctors' claims of professionalism. 'Indeed, retention of absolute control over technical procedures is clearly an absolute necessity for the survival of modern medical power' (Oakley, 1987, p. 46). The term 'technological imperative' was originally used by Fuchs (1968) to suggest that the addition of any new technology generates an increase in further use by its very existence, and this in turn generates still more technology.

There are a number of interlocking socio-economic factors which generate the development and use of medical technologies before their appropriateness and efficiency are determined, even before the grounds for their increased use are established (McKinlay, 1981). What are the dynamics of this process in accounting for the massive expansion of medical machinery? Technology is central to claims of professionalism and this has two important related aspects: having power in the doctor–patient relationship and having power within the profession. Let us turn to the doctor–patient relationship first.

The professional hierarchy means that doctors are regarded as experts who possess technical knowledge and skill that lay people don't have. The doctor–patient relationship is also often a class one, with a meeting between a middle-class, highly educated professional and a working-class patient. As well as being gendered, the relationship is often characterized by racial inequalities. Technology plays a major role in consolidating this distancing of the doctor from a necessarily passive patient, leading to the dehumanization of health care. The growing supremacy of technology in contemporary medical practice is not by any means confined to obstetrics, and both male and female patients find it an alienating experience.

In modern Western medicine, technological advances have transformed the methods of diagnosing illness, and these new methods have in turn altered the relationship between physician and patient.[9] The ubiquitous stethoscope has its origins in the doctor's wish to keep the patient at a distance, overlaid with the requirements of modesty as between men and women. According to the apocryphal story, the stethoscope was invented in 1816 by Laennec, during the examination of a young woman who had a baffling heart disorder. Restrained by the patient's youth and sex from placing his ear to her heart he recalled that sound travelled through solid bodies. From rolling some sheets of paper into a cylinder on this occasion he went on to construct the first wooden stethoscope (Reiser, 1978, p. 25). The human ear was supplanted by the stethoscope not because of any technical deficiency but because of prevailing social mores.[10]

Broadly speaking, since the nineteenth century there have been three stages in the historical development of the methods used to diagnose illness. Physicians have moved 'from direct communication with their patients' experiences, based upon a verbal technique of information gathering, to direct connection with the patients' bodies through techniques of physical examination, to indirect connection with both the experience and bodies of their patients through machines and technical experts' (Reiser, 1978, p. 227).

During the course of the twentieth century doctors have increasingly come to rely on technologically generated evidence at the expense of physical examination and history-taking. Machines inexorably direct the attention of both the doctor and the patient away from experiential or 'subjective' factors and towards the measurable aspects of illness. Moreover, Reiser argues that many of the modern diagnostic machines which have supplanted the more traditional manual methods and simple instruments are of little real value. The fact that they are so commonly used is not an indication of the reliability of the 'objective' evidence they produce but rather a result of doctors' insecurity and corresponding dependence on them. The skills involved in medicine may actually be declining as a result of this overdependence on technology as doctors become less willing to make independent clinical judgements based upon their own abilities and experience.

Obstetrics is a special case because the patients are uniformly women, they are generally not ill, and it is clearly an area where male doctors can have no personal experience of the 'condition' being treated. So their claims to expertise might appear tenuous to women. Oakley argues that technology is particularly attractive to obstetricians because techniques such as the stethoscope and foetal monitoring enable male doctors to claim to know more about women's bodies than the women themselves.

Once the technology is available women as patients may well want and expect high-technology treatment. This does not make women the passive victims of reproductive technologies and the male doctors who wield them. Within limits, women who are already advantaged in the social structure may even experience 'a sense of greater control and self-empowerment than they would have if left to "traditional" methods or "nature"' (Petchesky, 1987, p. 72).

However, the routine use of ultrasound imaging in pregnancy continues despite scepticism as to its medical benefits. Indeed the basic technique of ultrasound was not designed for obstetric purposes at all. Its origins date from attempts to detect submarines through sound-waves during the First World War. The subsequent development of ultrasound as a medical technology was as an offshoot of a major research project on acoustics at MIT financed by the US Navy (Yoxen, 1987). The concentration on pregnancy came several years after its use in other clinical diagnostic fields and the interest in foetal abnormality and rates of growth, with which it is now mainly associated, came even later. The procedure serves to discredit and then displace women's own experience of the progress of the foetus in favour of scientific data on the monitor. Some feminist critics fear that these techniques turn women into mere spectators of their own medically-managed pregnancy. As such they represent the ultimate appropriation by men of women's knowledge and expertise.

Within Western medicine, the high technology activities are not only a key to power at the level of doctor–patient relations, but also to power within the profession. Status, money and professional acclaim within the medical profession are distributed according to the technological sophistication of the speciality. To be seen to be developing and expanding high technology procedures signals success in the competition for scarce resources - as between specialists, between hospitals, and between individuals. 'Medical specialization and technological innovation have a special feature: they are parallel and interactive. Medical specialization leads to technological innovation; then, as a given technology is used, physicians and industrial designers collaborate to improve it. As it is defined, that process leads to ever more specialization and associated work and procedures' (Fagerhaugh et al., 1987, pp. 7–8).

The evolution of the new techniques of in-vitro fertilization and embryo transfer illustrate this process. On the face of it, the current enthusiasm for the new techniques seems curious. After all, in-vitro fertilization and embryo transfer have proceeded without much further work on establishing causes of infertility or improving other treatments. Given their low success rate, and the level of physical danger and psychological distress that accompany these new reproductive technologies, why the current concentration on in-vitro fertilization among infertility specialists? How does it happen that resources are allocated to this 'unsuccessful' technology?

Whilst it is true that new technologies generally have a high failure rate until perfected, it is also the case that 'many roads' are not taken in science. There is as yet no detailed description of the stages in the origination of these procedures and techniques. We might ask, with Edward Yoxen (1985, p. 143), what set of career choices led Edwards and Steptoe into their collaboration, or why their interpretation of the risk studies in animals was much less cautious than anyone else's, or why there are so few data on the effects of the drugs and invasive procedures used in in-vitro fertilization, or why there are so few data on the causes of infertility. Questions of inventive success and failure can be made sense of only by reference to the goals of the people involved.

Professional interests explain a great deal about the development of these techniques. Before the introduction of in-vitro fertilization and embryo transfer, the investigation and treatment of infertility had long been afforded low status in the medical hierarchy. Many of the procedures were carried out by general practitioners, as they required little special knowledge. Naomi Pfeffer (1987) argues that the new techniques of in-vitro fertilization and embryo transfer provide gynaecologists with an exciting, high-status area of research as well as a technically complex practice which only they can use. Status and substantial financial reward are to be had, as well as job satisfaction.

By 1982 in Britain, the Royal College of Obstetricians and Gynaecologists was already claiming that their fields had expanded so much that it should be divided into four sub-specialities. 'The pressures towards sub-specialization within gynaecology and obstetrics, then, constitute another incentive for medical personnel involved in the treatment of infertility to lay claim to new areas of expertise' (Pfeffer, 1987, p. 88). Official recognition of sub-specialization would attract financial support for training and research. In many ways therefore it is apparent that professional interests play a central role in determining the type and tempo of technological innovation in this area.

There are however wider economic forces at work. The commercial interests of the vast biotechnology industry are particularly influential. Much has been written about the 'new medical–industrial complex' and the way in which resources are systematically channelled into profitable areas that often have no connection with satisfying human needs. There is as yet little detailed information about the financial interests of medical biotechnology corporations in the development of the new reproductive technologies. Furthermore the potential commercial applications of the products are as yet unclear, at least to the general public.[11] What is clear is that the needs of infertile women play only a small part in the research agenda envisaged. Embryos are a unique source of information about human genetics, embryonic development and foetal growth. As Professor Robert Winston, of the Hammersmith Hospital, West London, explained: 'We think that in-vitro fertilization is merely the first step. In the long term the embryo could be removed for a few hours and then be replaced. This suggestion is not pie in the sky.' (quoted in *The Guardian*, 1 January 1989).

By March 1989, male embryos had been distinguished from female embryos within days of conception. The potential for genetic screening is now immense and with it the possibility for gene transplant experiments, known as 'gene therapy'. The ultimate aim, which has attracted vast research funding in North America, Europe and Japan is to unravel all the instructions contained in the human genetic code.

Some commentators have likened the scope of the biotechnology revolution to that of the microelectronic revolution, seeing it as the next technology-based phase of capitalist development.[12] Already we can see that the human body is caught up in commerce in new ways, with human organs such as kidneys, eyeballs, frozen foetuses and gametes being traded on the international market. Whatever one's position on the ethics of embryo research it is clear that it is always structured by relations of exploitation based on race, class and gender. The traffic in Korean foetuses for American military research into biological warfare is a case in point.

Although women are the prime targets of medical experimentation, reproductive technology cannot be analysed in terms of a patriarchal conspiracy. Instead a complex web of interests has been woven here – those of professional and capitalist interests overlaid with gender. It is more specifically to the operation of gender divisions that we now turn. The next section will examine the dynamics of a technology less recent and better documented than those referred to above. Nowhere are sexual relations more profoundly formative of a set of technologies than in the sphere of contraception.

The Sexual Relations of Contraceptive Technology

The perspectives from which most histories of fertility control are written are redolent of technological determinism. The conventional view shared by historians and demographers is that in pre-industrial societies women were the victims of their own fecundity.[13] There is a tendency to look back from our current 'Pill Era' and regard birth control as a nineteenth-century invention, representing the triumph of the progressive forces of technology over ignorance and prejudice. The Pill, a technical invention, is credited with enabling women for the first time to control their fertility, and the massive social changes for women that accompanied its introduction are attributed to it.[14]

It is assumed that earlier generations were prevented from practising birth control because they lacked the necessary technology. Many accounts of the history of birth control begin with the invention of the condom, arguing that it was only in the nineteenth century with the manufacture of rubber devices that effective contraception was made possible (McLaren, 1984, p. 5). On closer analysis it is apparent that the extent and openness with which birth control is practised, and the form it takes, is as much dependent on a society's attitude to sex, children and the status of women, as it is on effective technology. 'For the use of birth control requires a morality that permits the separation of sexual intercourse from procreation, and is related to the extent to which women are valued for roles other than those of wife and mother' (Greenwood and King, 1981, p. 169).[15] Birth control has always been a matter of social and political acceptability rather than of medicine and technology. Like childbirth, its prevention has always been subject to elaborate regulation and ritual.

In her book on birth control in America, Linda Gordon (1977) argues that social institutions and cultural values, rather than medical or technical considerations, have shaped modern contraceptive technology. Like most feminists, Gordon began with the premise that birth control represented the single most important contribution to the material basis of women's emancipation in the last century. However, she was quickly led to ask why the technology of contraception developed when it did, and why, in our generation, the invention of the Pill is seen as the key to liberation. For her, birth control was as much symptom as cause of larger social changes in the relations between the sexes and in the economic organization of society.

The ability to transcend biology was present in the earliest known societies. 'There is a prevalent myth, in our technological society, that birth-control technology came to us with modern medicine. This is far from the truth, as modern medicine did almost nothing, until the last twenty years, to improve on birth control devices that were literally more than a millennium old' (Gordon, 1977, p. 25). In fact, most of our present methods have had precursors in societies far less technologically sophisticated than ours. There is evidence from old medical texts and from anthropological studies that women have almost universally sought to control their fertility. Far from being invented by scientists or doctors, effective forms of birth control were devised and administered by women in nearly all ancient societies.

Reproductive knowledge and practice has always been part of women's folklore and culture. The relatively recent establishment of the male hegemony in medicine has obscured the existence of earlier methods that were more under women's control. Traditionally, knowledge about techniques for birth control, like remedies for other complaints, was developed and practised by wisewomen and midwives and handed down from generation to generation. A wide array of birth-control techniques were practised in the ancient world and in modern preindustrial societies including magic, herbal potions, infanticide, abortion, coitus interruptus, vaginal sponges, douches, and pessaries.

Not only did these techniques vary in their effectiveness, but they had very different implications for sexual relations. Some techniques are more amenable than others to being used independently and even secretly by women; some give full control to men; others are more likely to be used co-operatively. The point that needs to be emphasized is that women and men might have conflicting concerns and goals in mind when contemplating fertility control and these are reflected in the different techniques available. We will return to this point later.

Gordon argues that it is only by looking at this heritage of birth control customs that we can comprehend the emergence of the birth control movement, for that movement took its strength from women's understanding that traditional methods of fertility control were being suppressed. In particular, while abortion had hitherto been the subject of moral controversy, it was not until the nineteenth century that it was actually criminalized. These abortion laws were intended to eliminate doctors' rivals such as midwives and to undermine traditional forms of reproductive control. The result of the medical and legal intervention in this crucial form of birth control was a decline in women's ability to effectively limit their pregnancies. What was new in the nineteenth century, then, was not the technology to control fertility but the emergence of a political movement that campaigned for the right to use contraception.

However, reproductive self-determination for women was not the primary catalyst of the birth control movement. Equally important were the populationist movements inspired by Malthusian theories which sought population control as the cure for poverty.

During the twentieth century, contraception and to a lesser extent abortion have become respectable, and largely regulated by the medical profession. However the influence of population-control ideology is still central to modern birth control programmes. Since the 1950s, birth prevention has become a major international industry and it is linked with the politics of state intervention in population planning. Populationist ideology not scientific discovery was the catalyst for the major financial investment in research on birth-prevention methods and, according to Elkie Newman (1985), influenced the specific techniques which have become available. The technological prerequisites for the development of an oral hormonal contraceptive had existed by 1938 but popular morality and pro-natalist policies delayed its development until the late 1950s. According to Newman, it was the sudden and popular fear of a world population explosion which legitimated work on the Pill and resulted in family planning services becoming a major part of aid packages to the Third World.[16]

How then do we explain the emphasis on hormonal contraceptives and, by contrast, the heavy neglect of barrier methods, which are classed as old-fashioned? 'Considering how much time, money and energy is now spent on birth-control research, we might expect to be able to choose from among, say, ten different kinds of barrier method or perhaps a range of 'morning-after' methods. Instead, our options are confined to essentially two barrier methods, the various hormonal methods, a few IUDs and abortion techniques, and a small but increasing number of sterilization techniques' (Newman, 1985, p. 135).

Although the Pill is the most reliable method of contraception, it is associated with dangerous health risks and side effects for women. Nevertheless doctors favour the Pill because it helps to avoid the ethical dilemmas of dealing with unwanted pregnancy and abortion, and it requires a minimum of time and skill while keeping contraception firmly under their control. From the doctor's point of view, the fact that this method does not require many visits to the clinic, and does not need to be explained at great length to the patient are additional advantages. It is also important not to underestimate the significance of the Pill's profitability. It is economical to produce and market and needs to be taken daily, thus generating vast profits for the pharmaceutical industry that supplies it.

Apart from the corporate interests involved, most of the research into medical contraceptive methods is done by men on techniques for use by women. Interestingly, the incentive for the development of the condom was not birth control but rather men's need for protection from venereal disease. Given that women and men have different patterns of sexual behaviour, might not these differences be reflected in the design of contraceptive technologies? Indeed, Pollack (1985, p. 76) argues that these technologies 'are developed from a patriarchal perspective, emphasizing the sexual enjoyment of men and underestimating the costs to women. Male sexual pleasure is the most significant factor taken into account in the methods which become available, and in the ways in which contraceptives are used'.

Certainly men prefer methods that 'interfere' least with their experience of sex, even at the expense of women's health and enjoyment. The reluctance of heterosexual men to either wear condoms (even in the AIDS era) or to have a vasectomy indicates the primacy of their sexual feelings over the medical risks women are taking. While some medical techniques for men have been developed, the dangerous effects tend to be played down far less than is the case with female methods.[17] 'In fact men have been very reluctant to volunteer for experiments with male methods of any kind, just as they have been generally reluctant to be sterilized. The World Health Organization recently decided not to put much money into research on male methods in the future - they simply cannot persuade enough men to try them!' (Newman, 1985, pp. 141–2).

The Pill is also the technology favoured by women. As women still have the prime responsibility for pregnancy, the Pill is chosen for its high degree of protection and for the control that women can exercise over its use. Many women feel uneasy about touching parts of their own bodies and this is often linked to their anxieties about sexual activity. Using this method does not involve touching one's genitals, does not require male cooperation or even knowledge, and it allows for 'spontaneous' sex. The Pill has the additional psychological advantage of separating contraception from sexual activity, both in time and anatomically. It does not interfere with what is considered to be 'normal' romantic heterosexual sex, that is, for men to be lustful and assertive and for women merely to surrender. By comparison, the fitting of caps or diaphragms does require some skill, and to use it one has to admit to a man and to oneself that one is planning to have sex.

The definition of sexual activity as heterosexual intercourse involving penetration provides the context in which contraceptives are researched, developed, distributed and used. Contraceptive methods are designed to fit in with male-defined sex. Freed from the responsibility and the practices involved in using the sheath or withdrawal, men have been able to concentrate more on their enjoyment of sex. For women too the Pill has meant more effective birth control which in turn has been translated into more possibilities of sexual pleasure for women. However, women's increased sexual independence has been at a high health cost. If the gains for women outweigh their losses it is because of the achievements of the women's movement and not the technology *per se*. The Pill has not brought about women's liberation; women have gained control over their lives through social and political mobilization.

The purpose of this section has been to suggest that sexual relations in combination with population policies and market forces have shaped contraceptive technology. And, in turn, the design or form of the technology has been crucial to its use. In order to understand why particular technologies have the effects they do, this chapter has provided an account of the context in which reproductive technologies have developed and diffused. We have seen the role that technology has had in the medicalization and mechanization of medicine in general and in the area of human reproduction in particular. While the overall effect of this has been the masculinization of an area that was previously a women's sphere, women who are already advantaged in society have been in a position to benefit from recent reproductive techniques. In this area as elsewhere, technologies operate within and reinforce pre-existing social inequalities.

NOTES

1. See Stanworth (1987, pp. 10–11) for a categorization of the various technologies that are grouped under the term 'reproductive technologies'.
2. As Michelle Stanworth pointed out to me, the pro-interventionist position is also endorsed by the historian Edward Shorter (1983) who shares with Firestone a belief that women are the victims of their bodies and that twentieth-century medical technology has released them to be equal to men.
3. According to a recent estimate (Rowland, 1988), the success rate of in-vitro fertilization schemes is only about 10 per cent.
4. Of course, the belief that the maternal instinct is normal does not apply to single women or to lesbians. The Warnock Committee in Britain (set up to advise the UK Government on reproductive technologies) for example recommended restricting such techniques as in-vitro

fertilization, egg donation, embryo donation and artificial insemination to stable, cohabiting heterosexual couples.

5 I would like to draw the reader's attention to McNeil, et al. (1990) which does locate reproductive technologies within the sociology of technology. The collection only came into my hands as this book went to press.

6 Feminists have also been concerned to expose the increase in hysterectomy, particularly for black and Third World women, as a form of involuntary sterilization or as 'a simple solution to everything from backaches to contraception' (Homans, 1985, p. 5).

7 See also Arney (1982) and Richards (1978).

8 Martin (1987, p. 48). Although this negative medical imagery pervades women's own images of their bodies, involving extreme fragmentation of the self, women also resist. By analysing the different speech women use according to their class and race, Martin provides examples of how women generate their own more self-respecting meanings for menstruation, menopause, and birth. Interestingly she discovered that white middle-class women were far more likely to uncritically accept the hegemonic medical model of their bodies than working-class women, black or white.

9 This account is drawn from Reiser (1978). See also Foucault (1973) and Jewson (1976).

10 This is not to deny that subsequently the stethoscope did become technically 'superior'.

11 It recently emerged that American insurance firms are taking a keen interest in the development of genetic screening for its potential use in the recruitment of employees.

12 In fact, Haraway (1985) and Yoxen (1986) argue that this biotechnology revolution is also a cultural revolution, in that the very meaning of life is being transformed. With the development of genetic engineering, the dominant image of nature becomes one of organisms as information-processing systems that can be reprogrammed. 'Thus our image of nature is coming more and more to emphasise human intervention through a process of design' (Yoxen, 1986, p. 30).

13 The classic study is Himes (1936).

14 Even Rosenberg (1979, p. 50), normally such an astute critic of technological determinism, falls into this trap: '. . . one might therefore well argue that the women's liberation movement is essentially due to the combination of declining fertility (in turn partly attributable to a more effective technology of contraception), on the one hand, and the electrification of the household chores, on the other. One need not be a technological determinist to argue that the social benefits of the new-found freedom of women in American society are, in large measure, the product of technological innovation.'

15 That the invention of reproductive technologies often long predate their widespread use is evidenced by, for example, artificial insemination. Although we think of this as a radical new means of separating

conception from sex, it was actually first performed in 1776 (McLaren, 1984, p. 13).

16 See also Doyal (1979), especially chapter 7 on 'Medicine and Imperialism'.

17 This point is well made in a parody of a new contraceptive technique for men reprinted in *Spare Rib* (Vol. 93, April 1980, p. 9) reproduced below.

SCROTAL INFECTION 'Only 2 died'

The newest development in male contraception was unveiled recently at the American Women's Surgical Symposium held at Ann Arbor Medical Centre. Dr Sophie Merkin, of the Merkin Clinic, announced the preliminary findings of a study conducted on 763 unsuspecting male undergraduate students at a large midwest university. In her report, Dr Merkin stated that the new contraceptive – the IPD – was a breakthrough in male contraception. It will be marketed under the trade-name 'Umbrelly'.

The IPD (intrapenile device) resembles a tiny folded umbrella which is inserted through the head of the penis and pushed into the scrotum with a plunger-like instrument. Occasionally there is perforation of the scrotum but this is disregarded since it is known that the male has few nerve endings in this area of his body. The underside of the umbrella contains a spermicidal jelly, hence the name 'Umbrelly'.

Experiments on 1,000 white whales from the continental shelf (whose sexual apparatus is said to be closest to man's) proved the umbrelly to be 100 per cent effective in preventing production of sperm, and eminently satisfactory to the female whale since it does not interfere with her rutting pleasure.

Dr Merkin declared the umbrelly to be statistically safe for the human male. She reported that of the 763 graduate students tested with the device only two died of scrotal infection, only 20 experienced swelling of the tissues. Three developed cancer of the testicles, and 13 were too depressed to have an erection. She stated that common complaints ranged from cramping and bleeding to acute abdominal pain. She emphasized that these symptoms were merely indications that the man's body had not yet adjusted to the device. Hopefully the symptoms would disappear within a year.

One complication caused by the IPD and briefly mentioned by Dr Merkin was the incidence of massive scrotal infection necessitating the surgical removal of the testicles. 'But this is a rare case,' said Merkin, 'too rare to be statistically important.' She and other distinguished members of the Women's College of Surgeons agreed that the benefits far outweighed the risk to any individual man. (reprinted from *East Bay Men's Centre Newsletter* and *The Periodical Lunch*, Ann Arbor, Michigan.)

4
Domestic Technology: Labour-saving or Enslaving?

> Out there in the land of household work there are small industrial plants which sit idle for the better part of every working day; there are expensive pieces of highly mechanized equipment which only get used once or twice a month; there are consumption units which weekly trundle out to their markets to buy 8 ounces of this nonperishable product and 12 ounces of that one. There are also workers who do not have job descriptions, time clocks, or even paychecks.
>
> Cowan, *From Virginia Dare to Virginia Slims*

The introduction of technology into the home has especially affected women's lives and the work that goes on in the household. Indeed, it has been suggested that we should conceive of an industrial revolution as having occurred in the home too, that 'the change from the laundry tub to the washing machine is no less profound than the change from the hand loom to the power loom' (Cowan, 1976, pp. 8–9). Women's unpaid work in the home, servicing men, children and others, has for a long time been seen by feminists as the key to women's oppression. Relieving women of this burden has been a major project of feminism. As in other spheres, considerable optimism has attached to the possibility that technology may provide the solution to gender inequality in the home.

Since the 1970s housework has finally become the object of serious academic study by historians, sociologists and even a few economists. This was part of a general concern with the relationship between the changing structures of industrial capitalism and the shaping of everyday life within the household. *The Sociology of Housework* by Ann Oakley published in 1974 marked an important break in treating housework as work within the framework of industrial sociology. In the same year, Joann Vanek's article on 'Time Spent in Housework' compared the findings of the US time use studies of housework from the 1920s to the late 1960s. She argued that the aggregate time spent on housework by full-time housewives had remained remarkably constant throughout the period, although there had been some redistribution of time between individual tasks. Her surprising conclusion, that the introduction of domestic technology had practically no effect on the aggregate time spent on housework, soon became the orthodoxy amongst feminists working in the area.

In recent years feminist scholars in North America, Britain and Australia have produced excellent material on the history of housework and domestic technology.[1] Considerable attention has been devoted to countering the myth that housework is the creation of discontented housewives in that it 'expands to fill the time available'. Feminists (commonly quoting Vanek's data) have emphasized that women's household tasks have not decreased with so-called 'labour-saving' appliances. Much of this literature has pointed to the contradictions inherent in attempts to mechanize the home and standardize domestic production. Such attempts have foundered on the nature of housework – privatized, decentralized and labour-intensive. Thus, in the words of one writer on the subject, 'substantial changes in household technology left the sex, hours, efficiency, and status of the household worker essentially unaltered' (McGaw, 1982, p. 814). Illuminating as these historical works are, few of them have examined the extent to which the technological changes described were the result of economic imperatives or individual choice. Neither have they questioned whether women welcomed or resisted these innovations.

The economic and social significance of the household has also been analysed from a different perspective, by 'post-industrial society' theorists (Gershuny, 1983, 1985; Toffler, 1980), whose concern about 'de-industrialization' has led some to see the household replacing the factory as the centre of social and economic life. Typically, they see technological change as the major cause of shifts in the provision for needs between the informal household economy and the formal economy. As I shall demonstrate, arguments based on technological determinism play a central role in both this and feminist streams of thought.

This chapter begins by looking at feminist material on the mechanization of housework and then considers the work of post-industrialists on the impact of technological innovation in the home. By way of challenge to these latter theorists, I go on to examine some alternative approaches to individualized housework. In the last section I look at some of the key factors operating in the design process of specific household appliances. My aim is to explore the way the design and promotion of domestic technologies have been shaped by existing ideologies of gender.

Industrialization of the Home and Creation of the Housewife

What was the relationship between the technological developments in the economy and those in the home? To what extent did new technologies 'industrialize' the home and transform domestic labour? Why, despite massive technological changes in the home, such as running water, gas and electric cookers, central heating, washing machines, refrigerators, do studies show that household work in the industrialized countries still accounts for approximately half of the total working time (Sirageldin, 1969)?

The conventional wisdom is that the forces of technological change and the growth of the market economy have progressively absorbed much of the household's role in production. The classic formulation of this position is to be found in Talcott Parsons' (1956) functionalist sociology of the family. He argues that industrialization removed many functions from the family system, until all that remains is consumption. For Parsons, the wife - mother function is the primary socialization of children and the stabilization of the adult personality; it thus becomes mainly expressive or psychological, as compared with the instrumental male world of 'real' work. More generally, modern technology is seen as having either eliminated or made less arduous almost all women's former household work, thus freeing women to enter the labour force. To most commentators, the history of housework is the story of its elimination.

Although it is true that industrialization transformed households, the major changes in the pattern of household work during this period were not those that the traditional model predicts. Ruth Schwartz Cowan (1983), in her celebrated American study of the development of household technology between 1860 and 1960, argued exactly that.[2] For her, the view that the household has passed from being a unit of production to a unit of consumption, with the attendant assumption that women have nothing left to do at home, is grossly misleading. Rather, the processes by which the American home became industrialized were much more complex and heterogeneous than this.

Cowan provides the following explanations for the failure of the 'industrial revolution in the home' to ease or eliminate household tasks. Mechanization gave rise to a whole range of new tasks which, although not as physically demanding, were as time consuming as the jobs they had replaced. The loss of servants meant that even middle-class housewives had to do all the housework themselves. Further, although domestic technology did raise the productivity of housework, it was accompanied by rising expectations of the housewife's role which generated more domestic work for women. Finally, mechanization has only had a limited effect on housework because it has taken place within the context of the privatized, single-family household.

It is important to distinguish between different phases of industrialization that involved different technologies. Cowan characterizes twentieth-century technology as consisting of eight interlocking systems: food, clothing, health care, transportation, water, gas, electricity, and petroleum products. While some technological systems do fit the model of a shift from production to consumption, others do not.

Food, clothing, and health care systems do fit the 'production to consumption' model. By the beginning of the twentieth century, the purchasing of processed foods and ready-made clothes instead of home production was becoming common. Somewhat later, the health-care system moved out of the household and into centralized institutions. These trends continued with increasing momentum during the first half of this century.

The transportation system and its relation to changing consumption patterns, however, exemplifies the shift in the other direction. During the nineteenth century, household goods were often delivered, mail-order catalogues were widespread and most people did not spend much time buying goods. With the advent of the motor car after the First World War, all this began to change. By 1930 the automobile had become the prime mode of transportation in the United States. Delivery services of all kinds began to disappear and the burden of providing transportation shifted from the seller to the buyer (Strasser, 1982). Meanwhile women gradually replaced men as the drivers of transport, more and more business converted to the 'self-service' concept, and households became increasingly dependent upon housewives to provide the service. The time spent on shopping tasks expanded until today the average time spent is eight hours per week, the equivalent of an entire working day.

In this way, households moved from the net consumption to the net production of transportation services, and housewives became the transporters of purchased goods rather than the receivers of them. The purchasing of goods provides a classic example of a task that is generally either ignored altogether or considered as 'not work', in spite of the time, energy and skill required, and its essential role in the national economy.

In charting the historical development of the last four household systems, water, gas, electricity, and petroleum, Cowan reveals further deficiencies in the 'production to consumption' model. These technological changes totally reorganized housework yet their impact was ambiguous. On the one hand they radically increased the productivity of housewives: 'modern technology enabled the American housewife of 1950 to produce singlehandedly what her counterpart of 1850 needed a staff of three or four to produce; a middle-class standard of health and cleanliness' (1983, p. 100). On the other hand, while eliminating much drudgery, modern labour-saving devices did not reduce the necessity for time-consuming labour (see figure 4.1). Thus there is no simple cause and effect relation between the mechanization of homes and changes in the volume and nature of household work.

Indeed the disappearance of paid and unpaid servants (unmarried daughters, maiden aunts, grandparents and children fall into the latter category) as household workers, and the imposition of the entire job on the housewife herself, was arguably the most significant change. The proportion of servants to households in America dropped from 1 servant to every 15 households in 1900, down to 1 to 42 in 1950 (Cowan, 1983, p. 99). Most of this shrinkage took place during the 1920s. The disappearance of domestic servants stimulated the mechanization of homes, which in turn may have hastened the disappearance of servants (see figure 4.2).

This change in the structure of the household labour force was accompanied by a remodelled ideology of housewifery. The development in the early years of this century of the domestic science movement, the germ theory of disease and the idea of 'scientific motherhood', led to new exacting standards of housework and childcare.[3] As standards of personal and household cleanliness rose during the twentieth century women were expected to produce clean toilets, bathtubs and sinks. With the introduction of washing machines, laundering increased because of higher expectations of cleanliness. There was a major change in the importance attached to child rearing and mother's role. The average housewife had fewer children, but modern 'child-centred' approaches to parenting involved her in spending much more time and effort. These trends were exploited and further promoted by advertisers in their drive to expand the market for domestic appliances.

Housework began to be represented as an expression of the housewife's affection for her family. The split between public and private meant that the home was expected to provide a haven from the alienated, stressful technological order of the workplace and was

By means of the patent Electrolux attachment (here shown enlarged) corners are made easily accessible. Without any effort, every particle of dust and fluff is quickly removed and the corner left spotlessly clean.
Electrolux is the embodiment of modernism. It has to a large extent solved the maid problem by enabling many women to do their own house cleaning without effort and in a little time. There is no phase of cleaning for which it is not adapted.
It will thoroughly clean behind picture frames, picture rails, settees, and cushions with the same facility and thoroughness as it does rugs and floors. There are no troublesome attachments; nothing to get out of order; nothing to replace.
TELEPHONE FOR FREE DEMONSTRATION IN YOUR OWN HOME
Electrolux
The New Cleanness
Sydney, Newcastle, Brisbane, Melbourne, Ballarat, Geelong, Hobart, Adelaide and New Zealand

4.1 Advertisement from *The Home*, 1 April 1926

expected to provide entertainment, emotional support, and sexual gratification. The burden of satisfying these needs fell on the housewife.

With home and housework acquiring heightened emotional significance, it became impossible to rationalize household production along the lines of industrial production (Ravetz, 1965). Cowan graphically captures the completely 'irrational' use of technology and labour within the home, because of the dominance of single-family residences and the private ownership of correspondingly small-scale amenities. 'Several million American women cook supper each night in several million separate homes over several million stoves' (Cowan, 1979, p. 59). Domestic technology has thus been designed for use in single-family households by a lone and loving housewife. Far from liberating women from the home it has further ensnared them. This is not an inevitable, immutable situation, but one whose transformation depends on the transformation of gender relations.

The relationship between domestic technology and household labour thus provides a good illustration of the general problem of technological determinism, where technology is said to have resulted in social changes. The greatest influences on time spent on housework have in fact come from non-technological changes: the demise of domestic servants, changing standards of hygiene and childcare, as well as the ideology of the housewife and the symbolic importance of the home.[4]

Gender Specialization of Household Technology

If domestic technology has not directly reduced the time spent on housework, has it had any effect on the degree of gender specialization of household labour? Is the general relationship women and men have to technology itself a significant factor in determining the division of labour in the home?

Available evidence suggests that domestic technology has reinforced the traditional sexual division of labour between husbands and wives and locked women more firmly into their traditional roles.[5] Because technologies have been used to privatize work, they have cumulatively hindered a reallocation of household labour. Some household appliances may have been substituted for a more equal allocation of household labour, in particular reducing the amount of time men engage in housework.

The allocation of housework between men and women is in fact much the same in households where the wife is employed and those in which she is not. Husbands in all social classes do little housework. Where men do undertake housework, they usually perform non-routine tasks at intervals rather than continually, and frequently the work is outdoors. This is in marked contrast to women's housework, the dominant characteristic of which is that it is never complete.[6]

Task-specific technologies may develop in such a way that women

THE ALL-GAS HOME

FEW women really like housework, and it is "up to" the men of the house to install the labour-saving appliances so easily obtained by every householder using gas.

GAS COOKERS.—Many improvements are found on the modern gas cookers. Is your's up-to-date?

GAS COPPERS.—No stoking, no messy "cleaning up." Washing brought to the boil in half the time.

GAS IRONS.—No risk of shock—easily connected to any existing gas fitting.

BATH HEATERS.—Do you enjoy the inexpensive luxury of an ever-ready hot bath?

GAS FIRES.—The only heating method providing for REGULATED heat. Healthy. Never failing. Safe.

Inspect all makes of every type of gas appliance on one floor and in one visit at the new Showrooms and Demonstration Hall.

CORNER PITT AND BARLOW STREETS
(Opposite Central Railway Steps).

Write for attractively illustrated Catalogue—free and post free.

Free Cookery Demonstrations daily at 2.30, also Friday night at 7.30. Attend and obtain particulars of the Weekly Cookery Competition. Valuable prizes may be won by all amateur cooks.

4.2 'The All-Gas Home'

can take over tasks previously done by other family members. For example, Charles Thrall (1982) found that in families which had a garbage disposal unit, husbands and young children were significantly less involved in taking care of the garbage and wives were more likely to do it exclusively. Similarly with dishwashers, which are cited as one of the few appliances that do the job better and save time, husbands were less likely to help occasionally with the dishes.[7] 'In other words, new technologies may reduce the amount of time men engage in housework and increase the time spent by women, a finding which contradicts conventional wisdom' (Bose, et al., 1984, p. 78). Women have not been the prime beneficiaries of domestic technology.

Women's and men's relationship to domestic technology is a compound of their relationship to housework and their relationship to machines. Men's relationship to technology is defined differently to women's. Cultural notions of masculinity stress competence in the use and repair of machines. Machines are extensions of male power and signal men's control of the environment. Women can be users of machines, particularly those to do with housework, but this is not seen as a competence with technology. Women's use of machines, unlike men's, is not seen as a mark of their skill. Women's identity is not enhanced by their use of machines.

The household division of labour is reflected in the differential use of technologies, as Cockburn's (1985) study confirms. Few of the women in her sample used a hammer or screwdriver for more than hanging the occasional picture or mending the proverbial plug. Fewer still would use an electric drill or even a lawnmower, as 'men were proprietorial about these tools and the role that goes with them' (p. 219). Generally, women used utensils and implements - the dishwasher, vacuum cleaner, car - rather than tools. The skills necessary to handle these utensils and implements are no less than the male skills of their husbands. But, as Cynthia Cockburn points out, women cannot fix these utensils and implements when they go wrong and are therefore dependent on husbands or tradesmen, so that finally 'it is men on the whole who are in control of women's domestic machinery and domestic environment' (p. 220).

Technologies related to housework are not the only technologies to be found in the home. Indeed the extent to which the meanings and uses of domestic technologies have a gendered character is perhaps even more clearly demonstrated with regard to the technology of leisure. While for women the home is primarily defined as a sphere of work, for men it is a site of leisure, an escape from the world of paid work. This sexual division of domestic activities is read onto the artefacts themselves.

For example, television viewing reflects existing structures of power and authority relations between household members. In a study of white, working-class nuclear families in London, David Morley (1986) found that women and men gave constrasting accounts of their experience of television. Men prefer to watch television attentively, in silence, and without interruption; women it seems are only able to watch television distractedly and guiltily, because of their continuing sense of their domestic responsibilities. Male power was the ultimate determinant of programme choice on occasions of conflict. Moreover, in families who had a remote control panel, it was not regularly used by women. Typically, the control device was used almost exclusively by the father (or by the son, in the father's absence) and to some extent symbolized his domestic power.

Video recorders, like remote control panels, are the possessions of fathers and sons. In order to highlight the 'gender' of various household objects, Ann Gray (1987) asked women to imagine pieces of domestic equipment as coloured either pink or blue. Although the uniformly pink irons and blue electric drills were predictable, the mixtures in between were revealing. Home entertainment technologies were not wholly a neutral lilac. Both the timer switch and the remote control switch of video recorders were deep blue, that is, used and controlled by men.

Women's estrangement from the video recorder is no simple matter of the technical difficulty of operating it. 'Although women routinely operate extremely sophisticated pieces of domestic technology, often requiring, in the first instance, the study and application of a manual of instructions, they often feel alienated from operating the VCR.' (Gray, 1987, p. 43) Rather, women's experience with the video has to be understood in terms of the 'gendering' of technology. When a new piece of technology arrives in the home it is already inscribed with gendered meanings and expectations. Assuming himself able to install and operate home equipment, the male of the household will quickly acquire the requisite knowledge. Along with television, the video is incorporated into the principally masculine domain of domestic leisure. Gray also points out, however, that some women may have developed what she calls a 'calculated ignorance' in relation to video, lest operating the machine should become yet another of the domestic tasks expected of them.

Technological Innovation and Housework Time

Attempts by 'post-industrial utopians' to conceive of the likely shape of the household in the future suffer from many of the intellectual defects that have misled analysts of domestic technology in the past. Much of the work of these theorists is speculative. The British economist and sociologist, Jonathan Gershuny (1978, 1983, 1985, 1988) has made the most sustained attempt to give empirical weight to post-industrial predictions about the household.

Gershuny's starting point has little in common with that of the feminist commentators. His work is directed at theories of post-industrial society which see the economy as being based increasingly on services rather than on manufacturing production. By contrast, Gershuny's main thesis is that the economy is moving toward the provision of services within the household, that is, to being a self-service economy.

Although not drawing on the feminist literature, Gershuny shares with it a recognition that unpaid domestic production is in fact work and takes it seriously as such. He goes on to argue for a reorientation of the way we study technical change. Instead of starting from the workplace, as is typical for example in economics, sociology and economic history, in his view we should start from the household.

Households have a certain range of needs, a set of 'service functions' that they wish to satisfy, such as 'food, shelter, domestic services, entertainment, transport, medicine, education, and, more distantly, government services, 'law and order' and defence' (1983, p. 1). Historically the means by which households satisfy these needs changes. Gershuny describes a shift from the purchase of final services (going to the cinema, travelling by train, sending washing to a commercial laundry) to the purchase of domestic technologies (buying a television, buying a car, buying a washing machine). A degree of unpaid domestic work is necessary in order to use such commodities to provide services. This model is used to explain the economic expansion of the developed economies in the 1950s and 1960s, which was based on the creation of new mass markets in consumer durables, electronics and motor vehicles. In this way domestic technology is of enormous economic significance, affecting the pattern of household expenditure, the industrial distribution of employment and the division of labour between paid and unpaid work.

Like Cowan, Gershuny argues that people make rational decisions in this area. However, whereas her emphasis is on moral values and the social nature of human desires and preferences, his emphasis is on prices. The household will choose between alternative technical means of provision on the basis of the household wage rate, the relative prices of final services and goods, and the amount of unpaid time necessary to use the goods to provide the service functions. However, Gershuny assumes that people have unchanging desires and respond to market signals, making narrowly economic decisions primarily in terms of prices but also in terms of domestic labour time per item. But human beings do change and the introduction of machines alters people's preferences and values. The main weakness in Gershuny's analysis is that he ignores the social and cultural dimensions of human desires.

Implicit in this analysis is the assumption that the household can be treated as a unity of interests, in which household members subordinate their individual goals to the pursuit of common household goals. Gershuny shies away from any attempt to explain decisions as to whether men or women should do domestic labour, instead simply referring to 'the traditional segregation of domestic tasks' and 'people's perception of their roles'. What this approach overlooks is that there are conflicts of interest between family members over the differential distribution of tasks and money, and this may well influence how decisions actually come about.

Let us see how this theory explains the widespread purchase and use of washing machines, as opposed to commercial laundries. Gershuny's account differs quite sharply from Cowan (1983, p. 110) who explicitly considers and rejects an economic-rationality argument on laundry. He argues that as the time needed to use a washing machine has fallen, and the price of washing machines relative to the price of laundry has fallen so their popularity has increased. These developments are not linear however. A central feature of Gershuny's model is that it predicts first a rise, then a plateau, and then a decline in the time spent on domestic labour.

The first phase constitutes the shift from the service to the goods – for example from commercial laundries to domestic washing machines. According to the model this is a rational decision because it is cheaper, even counting the housewife's labour. But clearly, the domestic time spent on laundry goes up at this point. And precisely because it is a cheaper form of washing clothes, it becomes rational to wash more clothes more often, to satisfy (high) marginal desires for clean laundry.

In the second phase, where washing machines are fairly widely diffused, competition between manufacturers at least partly takes the form of offering more efficient machines, replacing the twin tub with the automatic. At the same time, the desire for clean laundry will begin to stabilize – slowing the rate of growth in clothes to be washed. Hence, eventually, time spent in laundry will start to fall. Thus, Gershuny argues, an effect of this move to a self-service economy, is that the amount of time spent on housework has declined since 1960 (1983, p. 151).

Gershuny is so convinced that new technologies increase the productivity of domestic labour that, in a recent paper with Robinson (1988), he takes issue with the feminist 'constancy of housework' thesis. Whilst conceding that prior to the 1960s the time spent by women on domestic work did remain remarkably constant, he insists that a shift occurred at that point. Drawing on evidence from time-budget surveys in the USA and UK, as well as Canada, Holland, Denmark and Norway, he concludes that domestic work time for women has been declining since the 1960s, and even that men do a little more than previously. It is central to his argument that this is so, even after taking into account the effects of such socio-demographic changes as more women having paid jobs, more men being unemployed, and the decreasing size of families. Therefore the diffusion of domestic equipment into households must have had some effect in reducing domestic work time. As Gershuny comments elsewhere: 'it would seem perverse to refuse to ascribe a substantial part of this reduction to the diffusion of domestic technology' (1985, p. 151).

In fact on closer inspection, these findings are more in line with feminist theories about constancy of domestic work than the authors would lead us to believe. Although the central argument is that domestic work time has been declining for women between the 1960s and the 1980s, this is only the case with respect to 'routine' domestic work. Unpaid work is subdivided into three categories: routine domestic chores (cooking, cleaning, other regular housework), shopping and related travel, and childcare (caring for and playing with children).[8] While routine domestic work has declined, the time spent in childcare and shopping have substantially increased.

This finding, however, is entirely consistent with the feminist emphasis on the added time now devoted to shopping and childcare. Certainly the feminist concern with the constancy of housework has employed a broader notion that includes childcare and shopping. To argue that domestic labour time has reduced is only meaningful if it means that leisure or discretionary free time has increased. If however mechanization results in less physical work but more 'personal services' work in the sense of increased time and quality of childcare, then surely this does not mean a real decrease in work. I presume that Gershuny uses such a narrow definition of domestic work because of his interest in the impact of domestic technology. However, it is difficult to maintain that women's domestic work time has declined because of the diffusion of domestic equipment whilst arguing that men's domestic work time has marginally increased at the same time. Men's increase is explained in terms of changing norms and thus inadvertently Gershuny calls into question any direct connection between the domestic technology and the time spent on housework.

Indeed it seems that the preoccupation with increases in productivity due to technological innovation blinds many analysts to more fundamental social factors. For example, the presence or absence of children, their age and their number all have significantly greater effects on time spent in housework than any combination of technological developments. Similarly, the presence of men in a household increases women's domestic work time by at least a third. In contrast, for men, living with women means that they do less domestic work (Wyatt et al., 1985, p. 39). Furthermore, it has repeatedly been found that the amount of time women spend on housework is reduced in proportion to the amount of time they spend in paid employment.[9]

A major problem with most time-budget research is that it does not recognize that the essence of housework is to combine many things, usually concurrently. This has a profound bearing on the interpretation of time spent in childcare and the apparent growth of leisure time. For example, watching television or listening to the radio can be combined with childcare, cooking, ironing and washing laundry. And, as I have pointed out with the case of television, this data would be particularly revealing with regard to women. Time budgets do not analyse whether activities are undertaken exclusively or in combination with another activity. Perhaps, as Michael Bittman (1988) suggests, the private and gendered character of the household promotes the kinds of technological innovations that maximize the number of tasks that can be performed simultaneously. To resolve such issues we would need more detailed information about the extent of use of consumer durables, the material output of services performed in the home and the social significance that these activities have for people. Gershuny's focus on technological innovations and tasks *per se* seems indicative, once again, of a technicist orientation which sees the organization of the household as largely determined by machines.

A technicist orientation is also evident in much of the futuristic literature on 'home informatics'. Ian Miles (1988), who collaborated with Gershuny on the research into the self-service economy, has attempted to chart the next wave of technological innovations, the new information and communication technologies, and their effects on the household. He argues that the new consumer electronic products of the coming decade are of major economic and social significance. There is much speculation about the fully automated home of the future known as a 'smart house' or 'interactive home system', where appliances will be able to communicate with each other and to the house within an integrated system. Miles predicts that home informatics will bring substantial changes to people's ways of life, one of which will be to improve the quality of domestic work both in terms of the convenience and effort required. However, Miles (p. 134) gives no reasons whatsoever for his hope that this will result in 'the sexual redivision of labour between men and women in families'.

The sociological literature on the electronic, self-servicing home of the future remains remarkably insensitive to gender issues. In particular, it ignores the way in which the home means very different things for men and women. Many of the new information and communication technologies are being developed for the increasing trend towards home-centred leisure and entertainment. But leisure is deeply divided along the gender lines. Many of these technologies, such as the home computer, demand that the user spend considerable time and concentration mastering it. But women have a lot less time for play in the home than men and boys. Programming the electronic system for the 'smart house' may enhance men's domestic power. Furthermore, the possibilities of home-based commercial operations, from 'telebanking' and shopping to 'teleworking', are likely to involve more housework for women in catering for other home-based family members. Although Miles' subtitle is 'Information Technology and the Transformation of Everyday Life', what is striking about these new technologies is just how little power they have to transform everyday life within the domestic world.

Alternatives to Individualized Housework

Even the most forward looking of the futurists have us living in households which, in social rather than technological terms, resemble the households of today. A more radical approach would be to transform the social context in which domestic technology applies. In view of what has been said about the shortcomings of domestic technology, one is prompted to ask why so much energy and expertise has been devoted to the mechanization of housework in individual households rather than to its collectivization.

During the first few decades of this century there were a range of alternative approaches to housework being considered and experimented with. These included the development of commercial services, the establishment of alternative communities and co-operatives and the invention of different types of machinery. Perhaps the best known exponent of the socialization of domestic work was the nineteenth-century American feminist Charlotte Perkins Gilman. Rather than men and women sharing the housework, as some early feminists and utopian socialists advocated, she envisaged a completely professionalized system of housekeeping which would free women from the ties of cooking, cleaning and childcare.

The call for the socialization of domestic work was not unique to the early feminist movements. Revolutionary socialists such as Engels, Bebel and Kollontai also saw the socialization and collectivization of housework as a precondition for the emancipation of women. And they embraced the new forces of technology as making this possible. Writing in the 1880s, Bebel saw electricity as the great liberator: 'The small private kitchen is just like the workshop of the small master mechanic, a transition stage, an arrangement by which time, power and material are senselessly squandered and wasted' (1971, pp. 338–9). The socialization of the kitchen would expand to all other domestic work in a large-scale socialist economy.

The modern socialist states of Eastern Europe took up some of these ideas, establishing collective laundry systems in apartment blocks and communal eating facilities. Whilst these initiatives certainly represented a different use of technology, they did not challenge the sexual division of labour insofar as women remained responsible for the housework, albeit collectivised. These policies on domestic labour resulted from the economic necessity of drawing women into the workforce combined with the ideology of equality. It is still the case that communal eating places are used a great deal more in the German Democratic Republic than in the West, and in 1974 it was estimated that families who used these facilities saved nearly two and a half hours per day compared to families who did not (Kuhrig, 1978, p. 311). Saving time, however, is not the sole motive as the housing crisis and overcrowded living conditions also encourage this pattern.

History thus provides us with many examples of alternatives to the single-family residence and the private ownership of household tools. Why then, in the USA in particular, has the individualized household triumphed? In particular, why should women apparently be so complicit in a process that was so damaging to them?

> Shall we believe that millions upon millions of women, for five or six generations, have passively accepted a social system that was totally out of their control and totally contrary to their interest? Surely there must have been at least one or two good reasons that all those women actively chose, when choices were available to them, to reside in single-family dwellings, own their own household tools, and do their own housework. (Cowan, 1983, p. 148)

To argue that women just welcomed the new domestic technologies because they became available is to come perilously close to technological determination. On the other hand, how can women have consciously and freely chosen to embrace the new methods when they have been so discredited as a liberating force? It is tempting in these circumstances to see women as duped, as passive respondents to industrialization, and as victims of advertisers.[10]

Cowan argues that women embraced these new technologies because they made possible an increased material standard of living for substantially unchanged expenditure of the housewife's time. To this extent women were acting rationally in their own and their families' interests.

However, as the following passage illustrates, Cowan seems to find the most convincing explanation of the paths chosen in a set of values to which women subscribed – the 'privacy' and 'autonomy' of the family.

> . . . . when decisions have to be made about spending limited funds, most people will still opt for privacy and autonomy over technical efficiency and community interest . . . Americans have decided to live in apartment houses rather than apartment hotels because they believe that something critical to family life is lost when all meals are eaten in restaurants or all food is prepared by strangers; they have decided to buy washing machines rather than patronize commercial laundries because they prefer to wash their dirty linen at home . . . When given choices, in short, most Americans act so as to preserve family life and family autonomy. The single-family home and the private ownership of tools are social institutions that act to preserve and to enhance the privacy and autonomy of families.' (ibid., p. 150)

Cowan does here depict women as active agents of their own destiny rather than passive recipients of the process. However, an approach that gives such primacy to values and to the symbolic importance of the home inevitably plays down the material context of women's experience. It may be that the effectiveness of the professional experts in imposing new notions of domestic life on behalf of the ruling class has been overestimated, and the resistance they engendered ignored. Most of the available historical research is based on the rhetoric of the experts and ideologues rather than the reality of working-class women's lives. The domestic science movement was never as fully accepted as its advocates hoped (Reiger, 1985; 1986). The evidence rather suggests that women negotiated the ideology of housewifery and motherhood according to their actual circumstances and that major contradictions underlay this attempt to rationalize domestic life. Similarly, when the advertisers were playing on these ideological elements in marketing the new domestic products, women actively participated in accepting or rejecting this process.

In Britain in the 1930s, there already existed an 'infrastructure' for the communal provision of domestic appliances. There were municipal wash-houses and laundries, communal wash-houses in the old tenement blocks, and at this time several local authorities experimented with building blocks of flats, modelled on those built in Russia, and incorporating wash-houses, crèches and communal leisure areas. However, the communal provision of amenities was not always seen as progressive. It was associated in many people's minds with back-to-back houses with their shared water supply and sanitation, and a characteristic squalid view of rows of dustbins and WCs, and the tap at the end of the street. Interestingly, class differences emerged over this issue on the Women's Housing Sub-Committee, with some of the middle-class feminists on the committee more interested in the possibilities for communal childcare, laundries and other facilities. That working-class women favoured privacy and did not favour communal arrangements may have been based on their own experience of communal living in conditions of poverty.

It is important to recognize the extent to which individual choice is constrained by powerful structured forces. The available alternatives to single-family houses were extremely limited, especially for the working class. In fact, state policy in the area of housing and town planning played a key role in promoting privatism. Without the extensive provision of different options, it is not clear to what extent people freely chose private domestic arrangements.

It is even less clear to what extent women, as opposed to men, exercised the degree of choice available. Oddly Cowan separates this American preference for domestic autonomy from the sexual division of domestic labour. No role is granted to men in choosing this single-family home even though Cowan's own historical findings point to men being well served by the private domestic sphere.

The common feminist stress on the negative effects of domestic technology has contributed to the view that women have been duped. There is a tendency among some feminist scholars to assume an unqualified anti-technological stance and to imply that modern housewives are worse off than their grandmothers (Reiger, 1986, p. 110). This tendency is evident in those authors who stress the increasing isolation of the domestic worker and see domestic labour as having lost much of its creativity and individuality. Once we recognize that the mechanization of the home did bring substantial improvements to women's domestic working conditions, even while it also introduced new pressures, women seem less irrational. 'When manufacturers then, in their own interests, marketed washing machines in terms of "make your automatic your clothes basket and wash every day", they were tapping into women's experience of the problems of organizing laundry and the physical drudgery it entailed. They were also opening up greater flexibility in managing some domestic tasks' (Reiger, 1986, pp. 115–16).

Against this there is no doubt that people can be taken in by false promises, especially where advanced technology is involved. Wanting to save time and improve the quality of their housework and in turn the quality of their home life, housewives are susceptible to well-targeted advertising about the capacity of new appliances to meet their needs. The irony is that women have commonly blamed themselves for the failure of technology to deliver them from domestic toil, rather than realizing that the defects lie in the design of technologies and the social relations within which they operate.

Men's Designs on Technology

Thus far my discussion of the literature on domestic technology reveals a preoccupation with its effects on the organization of the household and women's work in the home. However technologies are both socially constructed and society shaping. At a general level, I have argued that the predominance of the single-family household has profoundly structured the form of technology that has become available. There has been much less attention given to the innovation, development and diffusion processes of specific technologies themselves.

The forms of household equipment are almost always taken as

given, rather than being understood in their social and cultural context. Yet there are always technological alternatives and any specific machine is the result of non-technological as well as technological considerations. A society's choices among various possible directions of technological development are highly reflective of the patterns of political, social, and economic power in that society. Is it possible to detect these patterns in the design of domestic technology?

Gender relations are most obviously implicated in the development of domestic technology because of the extent to which the sexual division of labour is institutionalised. Most domestic technology is designed by men in their capacity as scientists and engineers, people remote from the domestic tasks involved, for use by women in their capacity as houseworkers. And, as we have seen, modern household equipment is designed and marketed to reinforce rather than challenge the existing household-family pattern.

It is not only gender relations that influence the structure of domestic technology. Like other technologies, domestic technology is big business. Particular technologies are produced not in relation to specific and objectively defined needs of individuals, but largely because they serve the interests of those who produce them. The design and manufacture of household appliances is carried out with a view to profit on the market. And the economic interests involved are not simply those of the manufacturers, but also those of the suppliers of the energy needed by these appliances.

Household appliances are part of technological systems, such as electricity supply networks. The interests of the owners of these systems have played an important part, along with those of the manufacturers, in shaping domestic technology. There is nothing the owner of an electricity supply system, for example, likes better than the widespread diffusion of an electricity-using household appliance that will be on at times of the day when the big industrial consumers are not using electricity. Residential appliances (including heating and cooling equipment) use about a third of the electricity generated in the US today; the refrigerator alone uses about seven per cent. Unlike most other household appliances, the refrigerator operates twenty-four hours a day throughout its life. In fact, many American kitchens now contain between 12 and 20 electric motors. Indeed the drive to motorize all household tasks – including brushing teeth, squeezing lemons and carving meat – is less a response to need than a reflection of the economic and technical capacity for making motors.[11]

The failure or survival, on the basis of vested interests, of some machines at the expense of others has profoundly affected the way our houses and kitchens are both constructed and experienced. This issue is raised in many feminist histories of housework and Cowan (1983, p. 128) presents a detailed example, that of 'the rivalry between the gas refrigerator (the machine that failed) and the electric refrigerator (the one that succeeded)'. There were initially designs for both and, indeed, until 1925 gas refrigerators were more widespread than the electric models. Cowan argues that electric refrigerators came to dominate the market as a result of deliberate corporate decisions about which machine would yield greater profit. The potential market for refrigerators, as well as the potential revenue for gas and electric utility companies, was enormous. Large corporations, like General Electric, with vast technical and financial resources, were in a position to choose which type of machine to develop. Not surprisingly, with interests in the entire electricity industry, General Electric decided to perfect the design of the electric refrigerator. The manufacturers of gas refrigerators, although they had a product with real advantages from the consumer's point of view, lacked the resources for developing and marketing their machine.

So the demise of the gas refrigerator was not the result of deficiencies in the machine itself; rather, it failed for social and economic reasons. And in this, it is structurally similar to the cases of many other abandoned devices intended for the household. This story illustrates that we have the household machines which we have, not because of their inherent technical superiority, nor simply because of consumer preference, but also because of their profitability to large companies. In this way economic relations shape domestic technology. 'By itself, the gas refrigerator would not have profoundly altered the dominant patterns of household work in the United States: but a reliable refrigerator, combined with a central vacuum-cleaning system, a household incinerator, a fireless cooker, a waterless toilet, and individually owned fertilizer-manufacturing plants would certainly have gone a long way to altering patterns of household expenditure and of municipal services' (Cowan, 1983, p. 144).

What is so original about Cowan's work is that she goes beyond a general account of technological change to present a concrete historical analysis of contingency in the evolution, design and development of a specific technology. She demonstrates the possibility of alternative machines and examines carefully the reasons for the path taken. However, it is disappointing that many of the wider concerns of her book disappear here. Her account is wholly in terms of the interests of, and the power play between, the companies producing the refrigerators, and the gender dimension is lost. Housewives here are relegated to the role of consumer – 'they bought electric refrigerators because they were cheaper'. Our understanding remains incomplete without research on design alternatives which shows how the form of the household, and the sexual division of labour within it, actively shape artefacts. We need much more work of this kind on what shaped these machines in the first place.[12]

An important dimension glossed over in the literature on the development of domestic equipment is the culture of engineering. After all, engineers do not simply follow the manufacturers' directives; they make decisions about design and the use of new technologies, playing an active role in defining what is technically possible. As I discuss in greater depth in chapter 6, the masculinity of the engineering world has a profound effect on the artefacts generated. This must be particularly true for the design of domestic technologies, most of which are so clearly designed with female users in mind.

When women have designed technological alternatives to time-consuming housework, little is heard of them. One such example is Gabe's innovative self-cleaning house (Zimmerman, 1983). Frances Gabe, an artist and inventor from Oregon spent 27 years building and perfecting the self-cleaning house. In effect, a warm water mist does the basic cleaning and the floors (with rugs removed) serve as the drains. Every detail has been considered. 'Clothes-freshener cupboards' and 'dish-washer cupboards' which wash and dry, relieve the tedium of stacking, hanging, folding, ironing and putting away. But the costs of the building (electricity and plumbing included) are no more than average since her system is not designed as a luxury item. Gabe was ridiculed for even attempting the impossible, but architects and builders now admit that her house is functional and attractive. One cannot help speculating that the development of an effective self-cleaning house has not been high on the agenda of male engineers.

Domestic Technology: A Commercial Afterthought

The fact is that much domestic technology has anyway not been specifically designed for household use but has its origins in very different spheres. Consumer products can very often be viewed as 'technology transfers' from the production processes in the formal economy to those in the domestic informal economy.

Typically, new products are at first too expensive for application to household activities; they are employed on a large scale by industry only, until continued innovation and economies of scale allow substantial reduction in costs or adaptation of technologies to household circumstances. Many domestic technologies were initially developed for commercial, industrial and even defence purposes and only later, as manufacturers sought to expand their markets, were they adapted for home use. Gas and electricity were available for industrial purposes and municipal lighting long before they were adapted for domestic use. The automatic washing machine, the vacuum cleaner and the refrigerator had wide commercial application before being scaled down for use in the home. Electric ranges were used in naval and commercial ships before being introduced to the domestic market. Microwave ovens are a direct descendant of military radar technology and were developed for food preparation in submarines by the US Navy.[13] They were first introduced to airlines, institutions and commercial premises before manufacturers turned their eyes to the domestic market.

Despite the lucrative market that it represents, the household is not usually the first area of application that is considered when new technologies are being developed. For this reason new domestic appliances are not always appropriate to the household work that they are supposed to perform nor are they necessarily the implements that would have been developed if the housewife had been considered first or indeed if she had had control of the processes of innovation.

It is no accident that most domestic technology originates from the commercial sector, nor that much of the equipment which ends up in the home is somewhat ineffectual. As an industrial designer I interviewed put it, why invest heavily in the design of domestic technology when there is no measure of productivity for housework as there is for industrial work? Commercial kitchens, for example, are simple and functional in design, much less cluttered with complicated gadgets and elaborate fittings than most home kitchens. Reliability is at a premium for commercial purchasers who are concerned to minimize their running costs both in terms of breakdowns and labour-time. By contrast, given that women's labour in the home is unpaid, the same economic considerations do not operate. Therefore, when producing for the homes market, manufacturers concentrated on cutting the costs of manufacturing techniques to enable them to sell reasonably cheap products. Much of the design effort is put into making appliances look attractive or impressively high-tech in the showroom – for example giving them an unnecessary array of buttons and flashing lights. In the case of dishwashers and washing machines, a multitude of cycles is provided although only one or two are generally used; vacuum cleaners have been given loud motors to impress people with their power. Far from being designed to accomplish a specific task, some appliances are designed expressly for sale as moderately priced gifts from husband to wife and in fact are rarely used. In these ways the inequalities between women and men, and the subordination of the private to the public sphere are reflected in the very design processes of domestic technology.

In tracing the history of various domestic appliances, Forty (1986) shows how manufacturers have designed their products to represent prevailing ideologies of hygiene and housework. Thus, in the 1930s and 1940s manufacturers styled appliances in forms reminiscent of factory or industrial equipment to emphasize the labour-saving efficiency which they claimed for their products. At that time, domestic equipment was still intended principally for use by servants. However such designs made housework look disturbingly like real work and in the 1950s, when many of the people who bought these appliances were actually working in factories, the physical appearance of appliances changed. A new kind of aesthetic for domestic appliances emerged which was discreet, smooth, and with the untidy, mechanical workings of the machine covered from view in grey or white boxes.[14] The now standard domestic style of domestic appliances '. . . suited the deceits and contradictions of housework well, for their appearance raised no comparisons with machine tools or office equipment and preserved the illusion that housework was an elevated and noble activity', of housework not being work (Forty, 1986, p. 219).

Throughout this chapter I have been examining the way in which the gender division of our society has affected technological change in the home. A crucial point is that the relationship between technological and social change is fundamentally indeterminate. The designers and promoters of a technology cannot completely predict or control its final uses. Technology may well lead a 'double life' ' . . . one which conforms to the intentions of designers and interests of power and another which contradicts them – proceeding behind the backs of their architects to yield unintended consequences and unanticipated possibilities'(Noble, 1984, p. 325).

A good illustration of how this double life might operate, and how women can actively subvert the original purposes of a technology, is provided by the diffusion of the telephone. In a study of the American history of the telephone, Claude Fischer (1988) shows that there was a generation-long mismatch between how the consumers used the telephone and how the industry men thought it should be used. Although sociability (phoning relatives and friends) was and still is the main use of the residential telephone, the telephone industry resisted such uses until the 1920s, condemning this use of the technology for 'trivial gossip'. Until that time the telephone was sold as a practical business and household tool. When the promoters of the telephone finally began to advertise its use for sociability, this was at least partly in response to subscribers' insistent and innovative uses of the technology for personal conversation.

Fischer explains this time lag in the industry's attitude toward sociability in terms of the cultural 'mind-set' of the telephone men. The people who developed, built, and marketed telephone systems were predominantly telegraph men. They therefore assumed that the telephone's main use would be to directly replicate that of the parent technology, the telegraph. In this context, people in the industry reasonably considered telephone 'visiting' to be an abuse or trivialization of the service. It did not fit with their understandings of what the technology was supposed to be used for.

The issue of sociability was also tied up with gender. It was women in particular who were attracted to the telephone to reduce their loneliness and isolation and to free their time from unnecessary travel. When industry men criticized 'frivolous' conversation on the telephone, they almost always referred to the speaker as 'she'. A 1930s survey found that whereas men mainly wanted a telephone for business reasons, women ranked talking to kin and friends first (Fischer, 1988, p. 51).

Women's relationship to the telephone is still different to men's in that women use the telephone more because of their confinement at home with small children, because they have the responsibility for maintaining family and social relations and possibly because of their fear of crime in the streets (Rakow, 1988). A recent Australian survey concluded that 'ongoing telephone communication between female family members constitutes an important part of their support structure and contributes significantly to their sense of well-being, security, stability, and self-esteem' (Moyal, 1989, p. 12). The telephone has increased women's access to each other and the outside world. In this way the telephone may well have improved the quality of women's home lives more than many other domestic technologies.[15]

Conclusion: More Work for Social Scientists?

I started this chapter by noting how belated has been the interest in domestic technology and household relations. There is now a substantial body of literature on the history of housework and the division of labour in the home. In recent years too there has been growing interest in domestic technology both among feminist theorists and, from a different perspective, among post-industrial society theorists. This work is still relatively underdeveloped and much of the literature shares a technicist orientation whether optimistic or pessimistic in outlook. Technology is commonly portrayed as the prime mover in social change, carrying people in its wake, for better or worse. But history is littered with examples of alternative ways of organizing housework and with alternative designs for machines we now take for granted. In retrieving these lost options from obscurity the centrality of people's actions and choices is highlighted and with them the social shaping of technology that furnishes our lives.

An adequate analysis of the social shaping of domestic technology cannot be conducted only at the level of the design of individual technologies. The significance of domestic technology lies in its location at the interface of public and private worlds. The fact that men in the public sphere of industry, invention and commerce design and produce technology for use by women in the private domestic sphere, reflects and embodies a complex web of patriarchal and capitalist relations. Although mechanization has transformed the home, it has not liberated women from domestic drudgery in any straightforward way. Time budget research leaves us wondering whether technology has led to more flexibility in housework or to its intensification. To further our understanding of these issues we need more qualitative research on how people organize housework and use technology in a variety of household forms. Such research should distinguish between different types of domestic technology and examine the significance of gender in people's affinity with technology. Finally, the designers of domestic technology themselves have so far been subjected to very little investigation; an examination of their backgrounds, interests, and motivation may shed light on the development of particular products. By refusing to take technologies for granted we help to make visible the relations of structural inequality that give rise to them.

This portrait of domestic technology is certainly incomplete. In this chapter I have concentrated on domestic technology as a set of physical objects or artefacts and argued that gendered meanings are encoded in the design process. This process involves not only specifying the user but also the appropriate location of technologies within the house. For example, domestic appliances 'belong' in the kitchen, along with women, and communications technology such as the television are found in the 'family room'. This signals the way in which the physical form and spatial arrangement of housing itself expresses assumptions about the nature of domestic life - an issue to be taken up in the next chapter.

NOTES

1 Ravetz (1965) is one of the earliest articles inquiring into the historical impact of domestic technology on housework. For detailed references, see Cowan's (1983) bibliographic essays at the back of the book and McNeil (1987, pp. 229–30). See also Bose et al. (1984), for a comprehensive review of the contemporary research. As they point out, this research is limited by its focus on the 'ideal' white middle-class family, and contains virtually no evidence on variations across class and ethnic groups; neither does it encompass single-parent households or people living alone. The data is also limited by its failure to reflect different stages of the life cycle. A similar problem exists with much of the historical literature, as McGaw (1982, p. 813) notes. This has led many authors to exaggerate the rate of diffusion of domestic devices.

2 See also Ruth Schwartz Cowan (1976 and 1979).

3 There is now quite an extensive feminist literature on the domestic science movement and its attempt to elevate the status of housekeeping. See, for example, Ehrenreich and English (1975,1979) and Margolis (1985) on America; Davidoff (1976) and Arnold and Burr (1985) on Britain; and Reiger (1985) for Australia. Reiger's book, *The Disenchantment of the Home*, is the most interesting sociologically as she attempts to combine a feminist analysis of the role of the professional and technical experts of the period with a critique of instrumental reason. The infant welfare and domestic science movements are seen as being part of a general extension of 'technical rationality' in the modern world.

4 I am only referring to domestic technology here, as clearly medical technology is central to demographic changes in life expectancy and to birth control.

5 See Bose et al. (1984), Rothschild (1983), and Thrall (1982).

6 In my own qualitative study (1983) in a small market town in Norfolk, England, I found that men always did the 'outdoor' jobs - mowing the lawn, gardening, fixing the car, household repairs and, to a lesser extent, painting and decorating. While the husbands did have a responsibility for performing certain household tasks, these had very different characteristics from those the women performed. Of course, this contrast is exaggerated and depends partly on conventional conceptions;

lawn-mowing, for instance, is just as continuous as window cleaning. Nevertheless, there is a general distinction which is reinforced by popular evaluations. Indeed, these evaluations are intrinsic to the domestic division of labour.

7 The microwave cooker is another interesting case where further research is needed to show whether it results in men being more prepared to take up some cooking activities or whether it increases expectations so that mothers cook separate meals for different members of the family at different times.

8 A fourth residual category, odd jobs, is not considered in the article.

9 This might lead one to expect that women in the paid labour force might use their income to substitute consumer durables for domestic labour. Surprisingly however women in employment have slightly less domestic equipment than full-time housewives. From an analysis of the Northampton household survey data, collected in 1987 as part of the British ESRC Social Change in Economic Life Initiative, Sara Horrell found that there were no significant differences in the ownership of consumer durables between working women and non-working women.

10 In her 1976 essay, Cowan has a tendency to adopt this latter position, seeing the corporate advertisers 'the ideologues of the 1920s' as the agents which encouraged American housewives literally to buy the mechanization of the home. The interest of appliance manufacturers in mass markets coincided exactly with the ideological preoccupations of the domestic science advisers, some of whom even entered into employment with appliance companies. According to W. and D. Andrews (1974), nineteenth-century American women, anxious to elevate their status, believed that technology was a powerful ally.

11 The Australian Consumer Association magazine, *Choice*, recently found that many appliances were useless and that a lot of jobs were better done manually. For example, they found that a simple manual citrus squeezer was overall better than many of the electric gadgets.

12 A notable exception is Hardyment's (1988) book on domestic inventions in Britain which documents a multitude of discarded designs, such as sewing machines, washing machines, ovens, irons, wringers, mangles and vacuum cleaners, invented and developed between 1850 and 1950. Unfortunately, the book contains little analysis of the forces which shaped their development. At one point, the author makes the intriguing argument that it was the small electric motor (introduced in the 1920s) more than any other invention which led to the development of domestic machinery along private rather than communal channels. But Hardyment concludes that 'the potential of any machine should lie in the mind of its user rather than its maker' (p. 199), echoing her earlier statement that women should seize the technological means to liberate themselves. It is disappointing that in a book devoted to the history of domestic machines so little attention is paid to the gender interests involved in their production.

13 This point is made by Megan Hicks, 'Microwave Ovens' (MSc dissertation, University of New South Wales, 1987).

14 One can only speculate as to whether covering up the mechanical workings of appliances assisted in alienating women from understanding these machines and how to mend them.

15 However, the unintended consequences of a technology are not always positive. The diffusion of the telephone has facilitated the electronic intrusion of pornography into the home. Not only are abusive and harassing telephone calls made largely by men to women, but new sexual services are being made available. The French post office's Minitel service, which is a small television screen linked to the telephone, has seen a massive 'pink message service' arise. When it was introduced over ten years ago, the Minitel system was intended to replace the telephone directory. Since then it has developed thousands of services, the most popular being pornographic conversations and sexual dating via the electronic mail. When complaints have been made the French post office claim that they can do nothing to censor hardcore pornography as it is part of private conversations. One wonders how this might affect gender relations in the home.

5
The Built Environment: Women's Place, Gendered Space

> Whether the private home is a free-standing house in Frank Lloyd Wright's Broadacre City or a high tower flat in Le Corbusier's Radiant City, domestic work has been treated as a private, sex-stereotyped activity, and most architects continue to design domestic work spaces for isolated female workers. Hayden, *The Grand Domestic Revolution*

In every culture and historical epoch, domestic architecture is uniquely revealing about prevailing social relations and norms of household organization. The design of houses is imbued with values and ideas that both reflect and exert tremendous influence over the patterns and quality of our lives. In this chapter I want to broaden the discussion of household technology to include the house as a technological construct, and the built environment more generally. The built environment is taken to mean '. . . our created surroundings, including homes, their arrangement in relation to one another, to public spaces, transport routes, workplaces and the layout of cities.' (Matrix, 1984, p. 1)

In what follows, I will be arguing that the built environment reflects and reinforces a domestic ideal which emphasizes the importance of the home as a woman's place and a man's haven. Sexual divisions are literally built into houses and indeed the whole structure of the urban system. Architecture and urban planning have orchestrated the separation between women and men, private and public, home and paid employment, consumption and production, reproduction and production, suburb and city. While people do not actually live according to these dichotomies, the widespread belief in them does influence decisions and have an impact on women's lives.

The focus of much feminist literature has been housework and the implications of technological developments within the home. Post-war sociology has chiefly considered housing as an aspect of the distribution and transmission of social wealth and privilege, that is, as an aspect of social stratification. At an economic level, housing is a commodity and central to the generation of capitalist profits. It is only recently that the structure and shape of the house itself has been subjected to feminist analysis.

The physical form of buildings is usually taken to be the inevitable result of technological and engineering advances, for example, concrete and steel gave us the high-rise tower block of modernist architecture. Changes in the interior design of dwellings are likewise explained in terms of mechanical innovations. A classic example can be found in explanations of the changing location of the kitchen, which is often attributed to the invention of the Rumford stove. This combined stove for cooking and heating eliminated odour and pollution and is said to be thus responsible for the movement of the kitchen from the basement or rear of the house to its centre.[1]

Certainly innovations in building materials, engineering methods and domestic technologies are of major importance and make possible the development of new architectural forms. However, as with other technologies, the design of the built environment is stamped with wider social and economic relations. Historians of architecture provide many instances of physical structures and arrangements that incorporate explicit or implicit political purposes. One such example is the wide Parisian boulevard designed by Baron Haussmann to permit the movement of troops and thus prevent any recurrence of street fighting of the kind that took place during the revolution of 1848.

Michel Foucault's discussion of Bentham's Panopticon, an all-seeing architectural form designed to keep prisoners under constant surveillance, is a vivid illustration of how a building can itself embody techniques of control. Prisons though are not the only buildings that can be designed to institutionalize patterns of power and order. The new IBM headquarters in Sydney is curiously reminiscent of the Panopticon. Its open-plan offices and clear glass internal walls are intended to give the appearance of a 'status-free environment'. Hierarchies seem to be dissolved where even managers' offices have glass walls and are located close to their staff. In fact of course, what these arrangements achieve is the possibility for increased surveillance of staff, who must feel watched even when they are not. In this sense, the glass itself does the looking. Like the Panopticon, then, the structure of the building ensures that control is largely achieved through self-discipline.

Whilst domestic architecture may not provide us with such stark examples of the extent to which buildings incorporate techniques of social control, that women are constrained in particular ways by the form of the family dwelling is certain. The house both symbolizes patriarchal relations, and gives concrete expression to them. In the first part of this chapter I will chart the development of the modern house. In so doing, I will be arguing that architectural changes in the domestic arena are not simply driven by technological advances but are about expectations of women and men, and in particular are about the domestication of women.

The Ideal Home

Victorian Values

By the middle of the nineteenth century the key foundations of the modern domestic ideal had already been laid. The sentimentalization of domestic life that occurred in the Victorian period in both England and America can be understood as an attempt to check what was perceived as the disintegration of values in a rapidly changing society by placing the home, symbolically, in direct opposition to the factory. Leonore Davidoff and Catherine Hall (1987) have shown how the period 1780–1850 saw the emergence of new conceptions of the home in Victorian English society. The 'creation of the middle-class home' involved the separation of home and workplace, and the identification of the home as a private place, in which family relationships were of primary importance. It was women's responsibility to create a home as unlike the world of business and industry as possible, a place that would be the centre of moral rectitude. In the responsibilities attributed to women in the home, it was the pursuit of beauty that was emphasized most strongly for the sake of its moral effects upon members of the household, especially the children. Located between the aristocracy and gentry on the one hand and wage labourers on the other, '[i]t was the middle ranks who erected the strictest boundaries between private and public space' (p. 359).

The home was seen as the most appropriate setting for women's lives, as a sanctuary from the rigours and corruption of the outside world. Victorian ideology perceived women and children as especially close to nature, much more so than men who could withstand the dangerous influences associated with supposedly unnatural city life, provided they had their retreat at home. The ideology of the home as a haven had its corollary in the idealization of the rural village as the proper setting for community life. What has been called 'the Beau Ideal' was epitomized by the Victorian villa in a garden suburb. As we shall see, this image has endured and been reproduced in twentieth-century suburbia.

A central theme of feminist theory has been the spatial separation of the public and private spheres, and the restriction of women to the latter. However, the public/private dichotomy cuts across the work/home distinction in many ways; for example, the 'private' home contains within it public and private spaces. As the house was turned into 'a home' during the Victorian period, the lay-out of the building was gradually transformed. There was marked spatial segregation of the sexes, both between husband and wife and master and servant. The servants' quarters were furthest from the front door - domestic service being concealed from the public parts of the house.

Distinct zones were created for different activities such as cooking, eating, washing, sleeping, and formal social functions, with public and private spaces clearly demarcated. One room at the front of the house, the parlour, was now set aside specifically for social intercourse and contained the best fittings and furniture. Eating now took place in a separate dining room. The kitchen was isolated at the back of the house and was spacious enough to accommodate several women working. 'Segregating the mess and smell of food preparation from the social ritual of eating became an important hallmark of respectability and meant that the kitchen became ideally as remote as possible from the living rooms, no matter the cost in servants', or wife's time and labour' (ibid., p. 383). Overall, then, hierarchy and location of rooms reflected the stratified relationships within the home - characterized by the subordination of servants to the family, family to wife and wife to husband.

The Home as a Machine for Living

Although in certain respects nineteenth-century ideas about what constituted a home have permeated twentieth-century life and thought, such as the physical and emotional division between public and private spheres, there have been major changes in our image of an ideal home. 'Most importantly, the nineteenth-century view of the home as a stronghold of beauty and spiritual virtue was replaced by the idea that the home's main function was as the source of physical welfare and health' (Forty, 1986, p. 114). Efficiency rather than beauty became the organizing principle of the home, and the relative importance attached to the various rooms changed, with the kitchen becoming the core of the house.

These changes reflected the growth of a middle class without servants and the mechanization of the home. The early years of the twentieth century saw the development of the domestic science movement, the germ theory of disease and the idea of 'scientific motherhood'. Standards of personal and household hygiene rose as did preoccupations with motherhood and children. The manufacturers and advertisers of domestic products exploited and promoted these attitudes in their drive to expand the market for domestic appliances. These appliances were promoted as labour saving and therefore a solution to the 'servant problem', supposedly enabling women to 'manage' their own labour 'scientifically'.

These ideas about efficient home management which accompanied the introduction of domestic technologies gradually reshaped the design of the house itself. Houses were being built for single family occupation without maid's rooms. Great emphasis was placed on the interior plan of the house in order to design efficient spaces to minimize the housekeeper's work. The idea that housework could be rationalized according to the principles of scientific management led architects to devote much time in the 1920s and 1930s to studying the logical sequence of work processes in the kitchen. Metaphors which described the kitchen as a laboratory prevailed. These ideas of functional, labour-saving homes became associated with the modern movement in architecture. Le Corbusier's famous phrase 'the home is a machine for living' captured this new view. While espousing emancipatory, indeed socialist-inspired politics, it seems the modernists did not appreciate that machines need constant servicing.

For women, the machine was to become a treadmill. The kitchen, now designed for the servantless family, was a compact fitted kitchen with room for one worker, the housewife. Neither its small size nor its location, sealed off from the rest of the house, were conducive to the sharing of kitchen duties. As with kitchens, internal bathrooms became standard in many households during this period because of the mass production of cast iron enamelware, as well as the obsession with dirt and disease. This model was to become the prototype for working-class as well as middle-class homes after the Second World War.

Public Housing for Private Lives

So far I have been concentrating on the development of a domestic ideal based on the family and on private life and its architectural embodiment in the single-family house. In the period between the wars most working-class families still lived, ate, cooked, and spent all their time in one main room, with shared facilities. Large-scale state intervention in the housing market in the aftermath of both world wars in Britain was to play a key role in ensuring that the single-family household became the dominant form of housing. Concerns about social improvement for the working-class and fears about social unrest were the twin motives of state reforms. As Gittins has argued: '. . . the state was actually *defining* what family structure and home life should be; the *size* of council houses, for instance, betrayed what the government felt to be the normative, the "right", size of family'. (Gittins, 1982, p. 48)

The debate about what sort of housing should be provided for the working class in the inter-war period was one in which women themselves participated. Set up in 1918, the Tudor Walters Committee, consisting exclusively of male politicians, architects and technical experts, was influenced by the architects associated with the Garden City Movement.[2] They recommended low-density terrace housing with wider frontages to provide more air and light, and ideally the living room was to be a 'through' room from the front to the back of the house. This room would maximize sun and ventilation in the house and emphasize family togetherness. The simplest and cheapest house model had a living-room with a range where most cooking would be done and a scullery with a gas cooker for occasional use, sink, copper and bath.

Working-class women interviewed at the time by members of the Women's Housing Sub-Committee insisted that a separate parlour be provided where they could relax and escape from work unfinished in the rest of the house. It has been said that: 'working-class women fiercely defended their right to a room which expressed their pride in housewifery and which also afforded additional privacy, a scarce commodity in working class households' (Lewis, 1984, p. 29).[3] However, given that maintaining the neat appearance of the parlour added to women's work, it is not entirely clear why working-class women are reported to have been such fierce advocates of its inclusion in their houses. The presence or absence of a parlour in working-class housing became quite a political issue in the inter-war period. Influential architects such as Raymond Unwin saw it as impractical to split the house into various small rooms and attacked the respectable working-class desire for a front parlour that was rarely used as 'a desire to imitate the middle-class house'. Where the architects' views prevailed, there are various stories of occupants dividing up the space themselves with partitions to create a separate parlour, thus demonstrating that working-class people resisted architects views of proper home life.

Housing the Symmetrical Family

If the Second World War posed a challenge to traditional sex roles, this was to leave little trace on the design of houses. In fact the post-war period saw the revitalization of the ideology of separate spheres for women and men. The major housing construction programmes of the 1950s and 1960s coincided with women being pushed back into the home to tend their husbands and children.[4] The housing stock which predominates today dates from or bears the stamp of this period. With the rapid growth of owner-occupation, greater emphasis was placed on a more home-centred lifestyle for men as well as women. The idea of companionate marriage saw the family as increasingly sharing activities and cultivating intimate relationships in the comfort of a private home. Here 'good' communication, intimacy, awareness of the needs of others, shared leisure (often shared consumption) gained a prominence previously accorded to hygiene and nutrition. But for all these apparent changes, the continuity with Victorian middle-class domestic ideals was in many ways more profound than the discontinuities.

This new socio-psychological conception of familial relations found its main expression in the open-plan housing design that characterized the post-war period. The dark divided house gave way to a preference for light and open space, breaking down traditional divisions between formality and informality in behaviour. Architects promoted the idea of multi-function spaces and 'zoned' planning in houses became the norm. Spaces were demarcated for certain functions, but this was achieved without separate rooms. The 'activity area' of the living room, dining area, and kitchen had few walls, providing as much space and togetherness as possible. The lack of walls was thought to promote the modern ideology of marital equality. Famous for its open-plan interiors, Frank Lloyd Wright's domestic architecture was nevertheless faithful to the Victorian iconography of family life by placing a massive hearth at the very centre of his house designs.

Domestic servants had finally completely bowed out of the home and consequently the illusion that meals simply arrived in the dining room – as if from nowhere – could no longer be sustained. There was therefore less reason to have a separate kitchen and dining room and the kitchen was now enlarged and opened up to the rest of the house. This open design gave domestic work a more egalitarian appearance, as other members of the family shared the space, and by implication the tasks, hitherto allocated to women alone. As we know, the domestic division of labour was not transformed by these architectural changes! However, they did obscure the extent to which women continued to bear responsibility for servicing the family. Typically, there was now a table in the kitchen for eating at, again signifying a less formal lifestyle. The open-plan kitchen enabled mothers to supervise children while cooking the meal, as children were now seen as requiring constant attention and companionship. This partially explains the move of the kitchen to the back of the house with a picture window looking out on the garden.

To cater for this increased concern with children's needs, the multipurpose room, which later became known as the 'family room', came into existence. 'Although the family room most often served as a place where children could do as they pleased in the midst of clutter and noise, it was also an architectural expression of family togetherness.' (Wright, 1981, p. 255) Very little privacy is provided for individuals within the house, which becomes primarily a place for shared activities. The bedrooms now provided for the children are generally small, ensuring that they will spend most of their time in the larger family room. Adults in the house are assumed to need even less private space – especially women. Even the parents' bedroom belongs to 'the Master'. Women do not have a room of their own, their spatial needs being subsumed into the family's: if they have a domain it is the kitchen.

The last twenty years have witnessed major shifts in the social position of women and in the way women see themselves. Paradoxically, this period has also been characterized by a renewed rhetoric about women as soft, feminine and housebound which is increasingly at odds with reality. The white plastic, clinical kitchen has given way to a more cosy 'country kitchen' with pine-panelled walls and natural wood finishes. Laura Ashley patterned floral prints recall the cheerful simplicity of rural life. Although most new houses now have central heating, the fireplace remains the focal point of many living rooms, with furniture grouped around it.[5] It is still the place of the most expensive furniture, with faint echoes of the Victorian parlour.

The kitchen meanwhile has become the emotional centre of the home: it is from here that the relaxed, informal, symmetrical family lifestyle radiates. Power relations within the patriarchal family have become submerged by this ideology of togetherness. Thus the prototype for the modern house prescribed the form of household that would inhabit it, namely the white middle-class nuclear family. As such it was not only oppressive to most women, but also a markedly ethnocentric design, denying the existence and needs of other forms of family. The dominant modern housing design does not lend itself to satisfying the housing needs of the majority of households today, which are in fact no longer composed of nuclear families.

Symbolic values about domestic life are perhaps even more clearly expressed in the external appearance of houses. The exterior of houses is the prime indicator of people's social status and extremely important to their self-image. Houses are, after all, the major article of consumption and their exterior is what counts most when they are purchased. Architects' prime concern has always been with the public face of buildings and the current debate on the nature of post-modern architecture is reproducing this concern. The contrast between domestic and commercial architecture is interesting in this regard. While non-residential architecture has gone through massive transformations in style, building materials, and construction technology, the preference for Georgian and Victorian domestic architecture remains. The facades of old houses are retained while the interior is gutted and modernized. There is even a market for new houses that are replicas of these styles, or in America of colonial-style houses.

While state-of-the-art commercial buildings pride themselves on being energy efficient and maintenance-free, the house still uses traditional materials such as wood and bricks that are both expensive and laborious to maintain. The assumption that women will continue to do much of this domestic work for free no doubt explains the disregard for efficiency in domestic architecture. However, the explanation is clearly more complex: men too are involved in maintaining the exterior of the home, investing much of their spare time and money in do-it-yourself home improvements. Furthermore, the preference for traditional architecture reflects an attachment to traditional values and a desire that the home should be a haven, resembling the workplace as little as possible. High-rise towers have met with little objection as offices but have proved very unpopular as homes.

Semi-Detached in the Suburbs

For all that privacy within the house has diminished, the expectation is that families as a whole remain private from each other. (Matrix, 1984, p. 55) The Victorian ideal of the detached or semi-detached house in a suburban or semi-rural setting remains essentially unchanged. The one-family house with a garden was regarded by the middle class and working class alike as the best place to bring up children, offering a healthy environment away from the dirt, noise and danger of the city. Developers encouraged the massive post-war move to the suburbs, as low-density development meant more profit for the building industry as well as providing a mass market for consumer durables. Although women have paid a heavy price for suburban development, they shared men's dreams of home ownership, their disillusion with the city and hopes for a better life in the suburbs. It took several decades for the aridness and uniformity of modern suburban life, and especially the isolation and boredom it forced on the housewife, to become immortalized in Betty Friedan's account of 'the problem with no name'.

Consonant with this idea of the home as private space, the distinctiveness of the home became enshrined in state zoning policies, which were at the heart of post-war town planning. Cities and towns were to be geographically segregated into their various activities, each with its appropriate location and setting. Zoning '. . . closely approximated stereotypical ideas about *man's* use of the environment' (Matrix, 1984, p. 38). It was assumed that the home and the neighbourhood were the setting for most women's lives and that men would travel to work located elsewhere. The main function of transport would be to get men from home to work and back again.

The impact that this would have on women's mobility was not considered. As Susan Saegert (1980) has observed, the long-standing symbolic dichotomy between 'masculine cities and feminine suburbs' fundamentally shaped the actual organization of the urban environment, tying women more closely to their immediate locality. Residential areas were and still are physically separated from industrial/commercial sites, distancing women from the 'economy'. Zoning thus intensified the privatized nature of many women's lives and their exclusion from the public, socially organized productive life. Suburban zoning restrictions have also operated to separate different sorts of housing development, limiting moderately priced high-density buildings to inner-city sites. As such it has been an important tool in class and race segregation – most infamously in South African urban planning, where black people are expressly confined to certain parts of the city.

Since at least the mid-1970s employers have responded to the separation of the workplace and home by relocating certain kinds of activities to the suburbs in order to capture the potential labour of married women who reside there. This has required the rezoning of some suburban space, especially in middle-class suburbs because it is white middle-class wives that are wanted for office work. Urban space is once more being restructured as the demand for clerical work expands - work traditionally done by women. Developments in information and communication technologies greatly facilitate large scale shifts in the nature and location of employment, and the decentralization of workplaces. It is not only office work that is now being 'suburbanized'. Industrial zones on the urban periphery have become massive centres of development. And suburban sprawl has stimulated the development of regional retail complexes. Meanwhile, administrative and financial activities - head office functions - remain located in the central city area. Overall this represents at least a partial shift away from mono-functional zoning to a more mixed use of urban space.

There is currently much interest in the contemporary restructuring of cities around 'service' sector work, and the rapid restructuring of manufacturing. Some of these analyses focus on the spatial constitution of power, that is, how the spatial allocation of goods, services, and employment across a city act as hidden mechanisms for the unequal distribution of income among various groups in the urban population. In Los Angeles, for example, it has been pointed out that industrial restructuring has left the largely Chicano/Hispanic and black industrial working class cut off from the new workplaces.

While such studies recognize the spatial construction of class and race differences, they generally ignore the issue of gender relations - aside from the obligatory listing of women with other disadvantaged or oppressed 'minority' groups.[6] There is little attempt to explore the different implications of such developments for women and men, and the ways in which the contemporary restructuring of cities affects the social relations of reproduction as well as the relations of production.

Feminist Alternatives: Would Women Do It Differently?

If the built environment tends to institutionalize patriarchal relations, is this because it has been designed and constructed predominantly by men? Would women, then, produce a different physical environment?

Planning and architecture in Britain, North America and Australia are indeed white, male-dominated professions. This is mirrored through all stages of building; even the production of the physical built structure is done by an almost all-male workforce. As the feminist designers' collective known as Matrix (1984, p. 3) comments, 'women play almost no part in making decisions about or in creating the environment. It is a *man-made* environment.' In their critiques of modern architecture, urban planning and of public/private distinctions, feminists have drawn attention to the sexual politics of space. They have suggested that the inevitable outcome of a profession and an industry inhabited and controlled by men is a male-defined built space.

The domestic architecture often cited as the epitome of a masculinist approach is the multi-storey residential block. This functionalist architecture, which envisaged a vertical garden city with 'streets in the air', has been discredited by feminists amongst others.[7] The fact that housework and childcare might be made more onerous and isolating for women stranded at dizzy heights, without safe and accessible outdoor space, did not occur to the pioneers of the Modern Movement. Apart from this obvious disregard for the quality of women's lives, these towers have also been seen as products of a specifically male vision (see figure 5.1). Modernism in architecture was obsessed with technological progress, adopting technology as both its instrument and symbol. The development of the high-rise form was a monument to technological innovation and a strikingly phallic symbol.

The underlying theme of such analyses is that women experience space differently from men and would therefore create different built environments. Margrit Kennedy, a Berlin-based architect, argues that 'there would be a significant difference between an environment shaped mainly by men and male values and an environment shaped mainly by women and female values' (1981, p. 76). Whereas men design a building from the outside in, women's greater preoccupation with interiors leads them to design buildings from the inside out. Kennedy suggests that there are the following male and female principles in architecture:

The Female Principles		*The Male Principles*
more user oriented	than	designer oriented
more ergonomic	than	large scale/monumental
more functional	than	formal
more flexible	than	fixed
more organically ordered	than	abstractly systematized
more holistic/complex	than	specialized/one-dimensional
more social	than	profit-oriented
more slowly growing	than	quickly constructed.

These ideas are echoed in many feminist critiques of architectural practice, which argue that whereas male subjectivity is expressed in tall phallic towers, female buildings are round, enclosing, curving and low-rise. Such views are not the prerogative of feminists alone.[8] In

5.1 Le Corbusier's theme of 'streets in the air' finds expression in Park Hill, Sheffield
Source: Architecture Department, University of Sydney

The City in History, Lewis Mumford proposed that in neolithic communities people lived in round dwellings, the house and the village being woman writ large: with the development of the city '[m]ale symbolisms and abstractions now become manifest: they show themselves in the insistent straight line, the rectangle, the firmly bounded geometric plan, the phallic tower and the obelisk . . .' (1961, p. 27).

Despite its initial appeal, there are a number of problems with this radical feminist position. To start with, the emphasis on universalized feminine and masculine traits in design cannot explain how it is that men as well as women have designed round and curving buildings. One need look no further than Gaudi's rippling architecture or the spiral shaped Guggenheim museum of Frank Lloyd Wright (see figure 5.2). Neither can it explain women's involvement in the design of high-rise buildings. As Kennedy herself remarks, in countries such as the USSR that have a high proportion of women architects, the dominant Western models of architecture prevail. Even though there are an increasing number of women practising architecture in Western

5.2 Interior of Guggenheim Museum, New York, by architect Frank Lloyd Wright
Source: Architecture Department, University of Sydney

countries, their professional education and training means that the work of women architects is not qualitatively different from that of male architects. That women architects have traditionally been assumed to be best suited for the design of domestic architecture and interiors reflects their low status in the profession rather than a specifically female attribute. It is to do with the hierarchical relationship between what is considered to be great 'architecture' of the public realm as opposed to the mere 'building' of houses.

On closer inspection several of Kennedy's characteristics of 'male' design are features of architecture operating within the constraints set by commercial imperatives. Women architects working under the same market pressures tend to design like men. To see central city office towers solely as the product of masculinist, phallocentric design values is to present a very partial picture which ignores investment calculations, capital flows, global property markets and the private ownership of land. The appearance of high-rise office buildings is explained as much by economic processes which lead to over-accumulated capital being invested in the central business district. As Margo Huxley (1988, p. 41) points out, these investments depend on 'political actions to retain the primacy of the central city and on the perceptions of (male) corporate directors of the prestige and power that is reflected in taking occupancy of the latest high-rise, high-tech office tower'.

While an account in terms of capitalist investment demonstrates the material basis of high towers, we still need an explanation of the cultural forces at work which give towers an association with power and prestige. I would argue that the cultural association between high-rise towers and male power is not only or primarily about their physical shape but is also because they represent the triumph of advanced technology. Perhaps this is why the radical feminist preference for low-rise 'human scale' development presents a credible if under-articulated alternative.

The risks inherent in the formulation of a specifically feminist architecture lie in the temptation to regard women as a homogeneous group. As Matrix (1984) emphasizes, there is a tendency to simply reflect the approach of white, middle-class women in the profession. Women's experience is very diverse, especially in terms of class. This is one of the interesting issues that is raised in Dolores Hayden's (1982) extensive research on nineteenth-century American feminist plans for utopian communities. Alternative approaches to individualized housework in single-family homes were proposed by an earlier women's movement. This 'lost feminist tradition' identified the economic exploitation of women's labour by men as the most basic cause of women's inequality. The central object of their campaigning was to socialize household labour and childcare. Most significantly, they sought to do this by a complete transformation of the spatial form and material culture of American homes, neighbourhoods and cities. Recognizing that the exploitation of women's labour by men was embodied in the actual design of houses, these 'material feminists' believed that changing the entire physical framework of houses and neighbourhoods was the only way to free women from domestic drudgery. They therefore urged architects and urban planners to explore radically new types of residential building.[9]

Two of the more influential women were Melusina Fay Pierce and Charlotte Perkins Gilman. In 1868 Melusina Fay Pierce, a middle-class Massachusetts woman, outlined plans for cooperative residential neighbourhoods made up of kitchenless houses and a cooperative housekeeping centre. She suggested that women organize to perform their household tasks cooperatively, building communal kitchens, laundries, dining facilities and childcare centres as necessary. Freed from the domestic routine, they would then be able to develop other interests outside the home. Writing in 1898, the economist Charlotte Perkins Gilman recommended kitchenless houses of a similar sort, suggesting that they be linked in urban rows or connected by covered walkways in a suburban block. Like Pierce, Gilman favoured the construction of kitchenless apartments with collective dining facilities for women with families. For Gilman however, the socialization of domestic work, rather than cooperation in its execution, was the means to economic independence for women. She envisaged a completely professionalized system of housekeeping which would free women from the ties of cooking, cleaning and childcare.

Ultimately this domestic reform movement foundered on the difficulty of overcoming both sex and class divisions in their urban and suburban communities. The problem of domestic service versus domestic cooperation could not be resolved. Many cooperative housekeeping societies accepted hierarchical organizational structures which put educated, middle-class managers at the top and paid dishwashers and laundry workers rather poorly. 'Feminists with capital who could afford the new physical environment for collective domestic work never thought of voluntarily sharing that domestic work themselves' (Hayden, 1982, p. 201). Thus, the liberation of professional middle-class feminists from domestic drudgery involved exploiting women of a lower economic class. The failure of this experiment in architectural solutions to the problem of women's domestic oppression is instructive. It demonstrates the impossibility of divorcing gender from class and other relations of inequality. It also demonstrates that new, egalitarian architectural forms cannot simply be superimposed on a preexisting social order and be transformative in themselves.

Automobiles: Technology in Motion

So far I have discussed the gender dimensions of housing design and urban layout. However, any discussion of the physical built environment is incomplete without discussing the transport technology that binds these spaces together. In particular the automobile is now a preeminent feature of the urban environment.

The invention and mass production of the car has greatly influenced the shape of the modern city. One has only to think of cities like Los Angeles and new planned towns like Milton Keynes, to be reminded of this. From the beginning of the modern movement in architecture, architects like Le Corbusier and Frank Lloyd Wright saw cars as integral to the design of the city. In this section I will argue that the transport system, and in particular the dominance of the car, restricts women's mobility and exacerbates women's confinement to the home and the immediate locality. Women's and men's daily lives trace very different patterns of time, space and movement, and the modern city is predicated on a mode of transport that reflects and is organized around men's interests, activities and desires, to the detriment of women.

The manufacture of automobiles is the largest industry in the world economy. It is dominated by a handful of American, Japanese and European companies that control 80 per cent of global production. In 1987, a record 126,000 cars rolled off assembly lines each working day, and close to 400 million vehicles are currently on the world's streets.[10] The automobile and its infrastructure dominate most North American and Australian cities in the literal sense that vast tracts of land are required to accommodate them. Not only for the motorways, but also for roundabouts, bridges, service stations, and parking spaces - at home, work, the supermarket and everywhere that people are supposed to congregate. Small wonder that in American cities, close to half of all urban space is dedicated to the automobile; in Los Angeles, the figure reaches two-thirds.

For the individual, the mobility and convenience that the private car bestows are unparalleled by any other means of transportation.

However, what appears to be an ideal solution to individual needs is increasingly illusory as more and more people choose, or are forced to make, similar decisions. In terms of individual mobility, the utility of the motor vehicle is diminishing as the number of cars on the road escalates. The prosperous 1950s and early 1960s were characterized by booming car ownership and, at least in the US and Australia, the car was expected to be the future of urban transport. The land use and transport planning procedures which emerged in the mid 1950s tended from the outset to be strongly associated with planning for roads and cars and pioneered the building of elaborate highway and freeway systems. However it transpired that freeways themselves spawned more and more traffic, becoming badly congested very soon after their completion. The obvious response to traffic congestion was to build more roads which were justified on technical grounds in terms of time, fuel and other perceived saving to the community from eliminating the congestion. 'This sets in motion a vicious circle or self-fulfilling prophecy of congestion, road building, sprawl, congestion and more road building.' (Newman, 1988, p. 15)

The net result is that London rush-hour traffic averages about 7 miles per hour; in Tokyo cars average 12 miles and in Paris 17. By comparison the average daily travel speed of 33 miles per hour in Southern California, where there are probably more miles of freeways than anywhere else in the world, may seem impressive. However as a result of a much lower population density than European cities, the advantage of speed is offset by the much longer distances required to travel to work. The irony is that a horse and buggy could cross downtown Los Angeles almost as fast in 1900 as an automobile can make this trip at 5pm today.

Motorway Madness

There is nothing inevitable about this rise and rise of the road. The state has played a major role in decisions about the extent to which transport investment is in roads as opposed to public transport. Again by comparison with Europe, American and Australian cities are characterized by a much heavier dependence on cars. Average Australian cities have four times, and US cities three times, more road supply per person than average European cities (Newman, 1988, p. 6). The politics of transport is dominated by conflict between road and rail lobbies, and technical discussions about efficient transport systems mask huge financial interests involved. The full extent of state subsidies to road transport are rarely exposed or documented. The expensive maintenance of motorways so heavily used by private road haulage companies is a case in point. The hidden subsidy to company car users via tax concessions and road maintenance is another. A Greater London Council study in 1986 found that the effective government contribution to company car users in London alone exceeded the revenue subsidy to London's public transport.

So far the story of the triumph of the car over other forms of transport technology may seem like another version of 'the paths not taken' argument, where people actively chose one type of technology in preference to others. However, in many cities that we now associate with the car, other forms of transport were not so long ago both preferred and extensively used. Contrary to popular impressions, Los Angeles is a sprawling metropolis not because of the automobile, but rather because it was built around the radial spurs of the electric railway system. It is almost forgotten today that in the United States there used to be a network of efficient and well-functioning urban and inter-urban rail systems in nearly every metropolitan area. By 1917, there were nearly 45,000 miles of trolley tracks which attracted billions of passengers. This transport system was not replaced by the motor car simply because of consumer choice. Rather, commercial interests joined forces at a key moment to close off all other options and ensure that henceforth investment would be channelled into automobile technology.

Beginning in the early 1930s, General Motors and other automobile tyre and oil interests, formed a holding company called National City Lines, whose sole objective was to purchase electric rail systems around the country and convert them to buses, which were manufactured and fuelled by members of the holding company. They acquired more than 100 rail systems in 45 cities, dismantled the electric lines and paved over the tracks. By the late fifties, about 90 per cent of the trolley network had been eliminated. The ultimate objective of this operation was to divert patrons of the earlier rail systems to General Motors cars. According to Snell, the reasons for this were clear: 'one subway car or electric rail car can take the place of from 50 to 100 automobiles'.[11] In 1949 General Motors, Standard Oil of California, and Firestone Tyres were found guilty of anti-trust conspiracy, but the damage had been done. By then, the political and economic power of the road lobby had succeeded in making American cities completely dependent upon the automobile. If there is a single force responsible for preventing the development of a diversified, balanced and ecologically-sound system of mass transportation, which was well within the bounds of the technologically feasible, it is the automotive and petroleum industries.

Women in the Slow Lane

If certain interests have conspired to make the motor car rule, the interests of certain social groups have been sacrificed to this end. The assumption of car ownership discriminates against the poor and the working class in general, and women constitute a disproportionate number of those affected. Older women and single mothers are among the poorest groups in society and have been literally left stranded in, or outside of, cities designed around the motor car. Although the automobile did not create suburbia, it certainly expanded and accelerated this process. The promotion of mass motor-car ownership has tended to exacerbate a greater dispersal of residential settlement often without any other mode of transport provided to service such areas.

These developments in transport policy have affected women and men differently.[12] Research on automobile use in Britain, North America and Australia indicates that proportionately more men than women have obtained drivers' licences, and that male car owners and drivers far outnumber female. Furthermore, while most women reside in car-owning households, evidence shows that women have considerably less access to the 'family car'. As a consequence of this, women are much more reliant than men on public transport to meet their travel requirements.

Despite women's low mobility, their travel needs are expanding as an increasing number of married women are entering the paid labour force and as the location of health care, educational resources and shopping facilities become more dispersed. Changes in patterns of consumption and service provision have increased the importance of transport access for women. For example, with the advent of the car, home-delivery services and corner stores gradually disappeared to be replaced by car-oriented supermarket complexes resulting in a significant increase in the proportion of time women spend on consumption activities. Even women who are not engaged in paid work must make frequent journeys to service the domestic needs of the household.

Although women are its primary users, in many ways public transport is not suitable for their needs and seems tailored to men's convenience. Recent work by geographers has drawn attention to the way the 'time-space maps' of the daily, weekly and overall life paths of individuals in their interactions with one another act as constraints on human activity. It has pointed to the major discrepancies between and within social communities in terms of fetters on mobility and communication. By emphasizing the critical connection between women's domestic roles and considerations of time and space, the time-geographic perspective adds a further dimension to our analysis of women's inequality. It has shown that the travel patterns of the two sexes are quite different and that, in response to domestic responsibilities, women elect to restrict the time spent on the journey to work. Given that family location is traditionally determined by its spatial relationship to the man's employment, women's opportunities are particularly restricted.

This is best illustrated by tracing the day-to-day activities of 'Jane', a single parent. 'Jane cannot leave home for work before a certain hour of the day because of her child's dependence on her for feeding and other needs, and because the sole accessible nursery is not yet open. Jane has no car and hence is faced with severe capability and coupling constraints in reaching the two 'stations' of the nursery and her place of work. Her choice of jobs is restricted by these constraints, and reciprocally the fact that she has little chance of acquiring or holding down a well-paid occupation reinforces the other constraints she faces in the trajectory of her path through the day. She has to collect her child in mid-afternoon, before the nursery closes, and is thus effectively restricted to part-time employment. Suppose she has a choice of two jobs, one better-paid and offering the chance to run a car, making it possible for her to take her child to a nursery further away from her home. On taking the more remunerative job, she finds that the time expended in driving to the nursery, to and from work and then back home again does not allow her time to do other necessary tasks, such as shopping, cooking and housework. She may therefore feel herself 'forced' to leave the job for a low-paid, part-time alternative nearer to home.'[13]

This exposition of a mother's day emphasizes the role played by transport facilities in constraining women's access to employment, services and social life. In particular, whether women are employed part-time, full-time or at all is to a significant extent contingent on these spatial relations. Firstly, an increasing number of women work part-time and therefore travel more in off-peak periods when services are more erratic. Yet public transport is still overwhelmingly designed around the needs of full-time workers commuting to the central business district. Secondly, as Jane's story demonstrates, women's journeys have been shown to be more complex or multi-purpose than men's as a result of their roles as mothers, unpaid domestic workers and paid workers. This means that they do many more journeys of shorter duration than men and these journeys are across the city. Even if the journey can be accomplished by public transport it requires a number of changes and is therefore very time consuming and expensive. This is a major reason why the job market for women is much more geographically restricted than that for men.

Furthermore, more women than men travel with grocery bags, baby carriages and dependants. Waiting at bus stops, climbing up and down bus steps or worse still underground stairs is a nightmare for anyone who isn't young, able bodied and unencumbered (see figure 5.3). The dominance of the car has also made the city an alienating environment for women and pedestrians. To get under motorways that divide cities requires passing through often dark, dingy underground passages where again there are often many steps to negotiate. 'Urban motorways and rural trunk roads cut through women's lives, driving a noisy, polluting, dangerous wedge between their homes and workplaces, schools and health centres, causing them to walk roundabout routes, through hostile subways or over windy bridges, diverting and lengthening bus journeys, and creating unsafe, no-go areas of blank walls and derelict spaces' (Women and Transport Forum, 1988, p. 121). Women are more vulnerable to sexual harassment and male violence while using or waiting for public transport. The Greater London Council's (1985) survey on women and transport discovered that nearly a third of women in London never go out alone after dark, and for Asian women the figure is 40 per cent. Of those who do travel at night, black and ethnic minority women feel less safe than do white women as they have the additional fear of racist attack. As public transport becomes more automated, there are fewer staff on trains, buses and platforms so women feel even more at risk. Interestingly the most car-dominated cities are the most dangerous. Detroit has one of the highest per capita murder rates of any city in the West. In cities like these cars are used as much for protection as for transportation.

I have been emphasizing the way in which the organization of the transportation system compounds women's inequality, virtually locking them into a world of very limited physical space, and exacerbates the unequal allocation of resources within the city. Perhaps the most revealing illustration of the way reliance on public transport can restrict the access of certain groups to public amenities comes from an article called 'Do Artifacts Have Politics?' by Langdon Winner (1980). Winner tells us that anyone who has travelled the highways of America and has become used to the normal height of overpasses may well find something a little odd about some of the bridges over the parkways on Long Island, New York. Many of the overpasses are extraordinarily low, having as little as nine feet of clearance at the

5.3 Urban planners ignore the needs of the least mobile
Source: Production stills from the film *Serious Undertakings* (1982) directed by Helen Grace. Photographs by Sandy Edwards

curb. Even those who notice this would not be inclined to attach any special meaning to it - we seldom give things like roads and bridges any consideration.

In fact, the two hundred low-hanging overpasses on Long Island were deliberately designed to achieve a particular social goal. Robert Moses, the master builder of roads, parks, bridges and other public works from the 1920s to the 1970s in New York, had these overpasses built to specifications that would discourage the presence of buses on his parkways. The reasons reflect Moses's class bias and racial prejudice. Affluent whites would be free to use their cars on the parkways for recreation and commuting. Poor people and blacks, who normally used public transport, were kept off the roads because the twelve-foot high buses could not get through the overpasses. One consequence was to limit access of racial minorities and low-income groups to Jones Beach, Moses's acclaimed public park. Although Winner does not mention women, women's dependence on public transport means that these physical arrangements also have a gender dimension.

This story illustrates that, far from being neutral, even seemingly innocuous technological forms such as roads and bridges embody and reinforce power relations. What is so significant about these vast technological projects is that they endure, such that for generations after Moses has gone, the highways and bridges he built to favour the use of the automobile over the development of mass transit continue to give New York much of its present form. 'Many of his monumental structures of concrete and steel embody a systematic social inequality, a way of engineering relationships among people that, after a time, becomes just another part of the landscape' (Winner, 1980, p. 124).

The Car Culture

Just as bridges may not be as innocent of political qualities as they may appear, so too cars have been shaped by a plethora of social and economic factors. Above I stressed that the dominance of the car was not simply about the efficient movement of people around cities but was ensured by economic forces. Means of travelling - whether by car, motorcycle or bicycle - are also consumer products charged with symbolic as well as economic and pragmatic meaning. The car is one of the central cultural commodities of the twentieth century: precisely because it is such a mass, commonplace technology, we often fail to appreciate its ideological significance. It is not simply technical efficiency that determines the design of cars but cultural forces that shape them.

Car manufacturers consciously design and style the appearance of their products to express consumer dreams, desires and aspirations. In turn, consumers purchase, along with their car, an image and a social identity. Cars are infused with powerful visual messages about the age, sex, race, social class and lifestyle of the user. Cars are a major feature of conspicuous consumption for men and have a central place in male culture. The masculine fantasies they represent take different forms, as can be seen by the contrasting designs of smooth, aerodynamic-style sports cars and the rugged, four-wheel-drive 'range rovers'. These have in common their symbolization of individual freedom and self-realization. Countless novels, films, popular songs and advertisements romanticize flight in a car and link cruising along the road with liberation. For men, cars afford a means of escape from domestic responsibilities, from family commitment, into a realm of private fantasy, autonomy and control.

Even more markedly than the car, the motorcycle is a symbolic object that represents physical toughness, virility, excitement, speed, danger and skill. Their conspicuous bodywork and mechanics resonate with their original military use, and speak of aggression and virility. Along with leather jackets, riders wear grease-stained jeans to express their technical competence. The experience of riding a bike encapsulates the outdoor, roving life of the wanderer with no ties. It also symbolizes a form of man's mastery of the machine; a powerful monster between his legs which he must tame. Trucks similarly are the giant iron horses of independent men who refer to themselves as 'cowboys' and boast of sexual encounters on the road. It is no accident that cars, trucks and motorcycles are usually personified as female and given women's names. They are after all the place where men feel most sexual, the vehicle for men's pursuit of sexual adventures, including their use of street prostitutes. In advertising their products, manufacturers associate these products with women's bodies and wild animals. Nubile women are draped over cars in advertisements. Men are the possessors and women the possessed. 'Manufacturers encourage the male user to perceive his machine as a temperamental woman who needs to be regularly maintained and pampered for high performance.' (Chambers, 1983, p. 308). Cars have long been a metaphor for sex and something wild in the already tamed urban environment. In recent years this imagery has become overlaid with new associations of the latest high tech computerization, bringing to the fore men's fascination with the power of technology – a theme further explored in the next chapter.

For all that I have been stressing that the car is a fetishized object for men, this is not the whole story nor the full extent of the gender relations embodied in the car. The design of the 'family car' reflects assumptions about the typical size of unit in which people wish to travel around. Furthermore, many cars are specifically designed with female drivers in mind. This is particularly explicit in the small hatchback car for 'running around town' and shopping. This is assumed to be the family's second car for the wife and mother to meet household needs. The powerful large car is destined for the male head of the household, although increasingly professional women are being targeted by manufacturers and advertisers as purchasers in their own right. Given the opportunity, women too enjoy driving fast and glamorous cars. However, for most women cars are a practical necessity to which they aspire for relief from drudgery and a release from home. They are also a relatively safe means of travel, given the violence and harassment to which women are subjected on public transport. And despite the prevalence of jokes about women drivers, in fact they are if anything more competent than men and much less likely to cause car accidents. Indeed the particular advantages that the car offers women sets up a tension for eco-feminism. While the car constitutes a major environmental hazard, for women, at least in the short term, demanding 'equal access' to the car is an important assertion of their right to independence, mobility and physical safety.

In this chapter I have been concerned to establish the connections between the built environment and patriarchy. The development of the modern house and the organization of domestic space within it reflects cultural assumptions about family relationships, the home as women's place and women's place being in the home. Sexual divisions are not only physically built into houses, but into the whole urban structure. The modern city is, furthermore, constructed around a mode of transport that reflects and is organized in men's interests to the detriment of women. Once we recognize the gendered nature of the design and production of the built environment, once it is no longer seen as fixed, we can begin to make space for women.

NOTES

1 According to W. and D. Andrews (1974, p. 316) the Rumford stove 'made possible the literal centralization of the woman to the activity of

the house; technology, in short, placed the woman in the midst of things and not removed from them'.

2 For details of the Tudor Walters Report, see J. Burnett (1978, pp. 218–221).

3 This quotation is taken from an excellent article by G. Allan and G. Crow (1990). For an account of the Women's Housing Sub-Committee, see B. McFarlane (1984) and A. Ravetz (1989).

4 Wright (1981, p. 256) notes that in the USA only 9 per cent of suburban women worked in 1950 compared with 27 per cent of the whole population.

5 To all intents and purposes of course the television is now the focal point of most living rooms, but it has not displaced the hearth, often being installed adjacent to the fireplace.

6 See D. Harvey (1989) and E. Soja (1989). One of the few articles that does attempt to draw out the implications of these changes for women is E. Harman (1983).

7 See A. Coleman (1985). For the classic critique, which is particularly interesting for its discussion of the consequences for bringing up children, see J. Jacobs (1962).

8 Indeed, some of these architectural principles have even gained royal approval! See The Prince of Wales (1989).

9 In England too there was much enthusiasm for cooperative housekeeping among the more socialist-inclined members of the Garden City Movement. Ebenezer Howard organized extensive experiments in cooperative housekeeping, building quadrangles of kitchenless units in the Garden Cities of Letchworth and Welwyn. According to Ravetz, however, the demise of domestic service was an important factor. 'It was perhaps this strong male interest in getting the housework done with minimum inconvenience to themselves that, more than any feminist inspiration, explains the interest of certain men or the garden city movement in collective housekeeping.' (Ravetz, 1989, p. 192).

10 The source of information for this paragraph is M. Renner (1988) and the *New Internationalist* No. 195, May 1989, issue on 'Car Chaos'.

11 The elimination of the interurban rail systems is documented in detail by B. Snell, 'Report on American Ground Transport', Subcommittee on Antitrust and Monopoly, Senate Judiciary Committee, 26 February 1974.

12 The study of women and transportation is quite new and some interesting themes are emerging. See M. Cichocki (1980) and S. Fava (1980); L. Pickup (1988); Women and Transport Forum (1988) and V. Scharff (1988).

13 This summary of R. Palm and A. Pred (1978) is taken from A. Giddens (1984, pp. 114–15). In the original article, the authors make the important point that the daily prisms of women in various stages of the life-cycle and in various social classes are different.

6
Technology as Masculine Culture

The link between technology and masculinity is commonly supposed to be self-evident and in no need of explanation. But as this chapter will show, the relationship, although strong, is more complex than may at first appear.

I have already argued that the traditional conception of technology is heavily weighted against women. We tend to think about technology in terms of industrial machinery and cars, for example, ignoring other technologies that affect most aspects of everyday life. The very definition of technology, in other words, has a male bias. This emphasis on technologies dominated by men conspires in turn to diminish the significance of women's technologies, such as horticulture, cooking and childcare, and so reproduces the stereotype of women as technologically ignorant and incapable. The enduring force of the identification between technology and manliness, therefore, is not inherent in biological sex difference. It is rather the result of the historical and cultural construction of gender.

This chapter will examine the ideological and cultural processes that serve to make 'natural', and thereby help to generate, this close connection between men and machines. That our present technical culture expresses and consolidates relations among men is an important factor in explaining the continuing exclusion of women. Indeed, as a result of these social practices, women may attach very different meanings and values to technology. To emphasize, as I do here, the ways in which the symbolic representation of technology is sharply gendered is not to deny that real differences do exist between women and men in relation to technology. Nor is it to imply that all men are technologically skilled or knowledgeable. Rather, as we shall see, it is the ideology of masculinity that has this intimate bond with technology.

In order to understand the underlying nexus between masculinity and technology, I will now consider several examples that highlight this identification, and where technology is seen as definitive of the activity in question. The chapter also examines women's relationship to masculine cultural worlds.

Men and Machines

Giving Birth to the Bomb: Virility and the Technology of Destruction

Contemporary feminist peace initiatives, like their forebears, have focused attention on the relationship between masculinity and war. Warfare is traditionally a male preserve and the connection between physical violence and men in both the public and private spheres is strong. So strong in fact that much feminist writing assumes that war, in the same way as rape, is the result of men's inherently aggressive nature. The weaponry of war is seen as intrinsically masculine so that cruise missiles have become a symbol of male power, the phallus. War is a paradigm of masculine practices because its pre-eminent valuation of violence and destruction resonates throughout other male relationships: relationships to other cultures, to the environment and, particularly, to women. Thus the threatened destruction of world civilization by nuclear arms is seen as the culmination of male developed and controlled science and technology. Many feminists have traced the source of the male fascination with weapons and war to biology and psychology, arguing that men need a substitute for the babies they cannot conceive.[1] Ironically, the most comprehensive account of this fundamentally radical feminist position is by a man, Brian Easlea, in his *Fathering the Unthinkable* (1983).

Drawing on the feminist analysis of the development of science as a form of domination of both nature and women, Easlea explains the creation of nuclear weapons in terms of the masculinity of science. He argues that 'the nuclear arms race is in large part underwritten by masculine behaviour in the pursuit and application of scientific inquiry' (p. 5). The book details the story of the discovery of radioactivity and the development of the atomic bomb and vividly describes the excitement and intense competitiveness of the prominent scientists who were involved. Easlea's purpose is to critique the kind of 'technical rationality' that dictated the making of the atomic bomb, reflected so clearly in J. Robert Oppenheimer's statement that 'when you see something that is technically sweet you go ahead and do it and you argue about what to do about it only after you have had your technical success' (p. 129).

What is striking is not only the compulsive nature of the work done by this group of men but the overriding pleasure and sheer joy they experienced in achieving technological perfection. 'Nobody worked less than 15, 16, 17 hours a day. There was nothing else in your life but this passion to get it done' (p. 84). They saw themselves as pioneers at the frontiers of what was only just possible, on a 'fantastic adventure' to being the first to release the awesome power locked in the nucleus. The Los Alamos physicists' reaction to the dropping of the bomb on Hiroshima makes particularly chilling reading. Prominent scientists recalled the exultation, celebration and pride they felt in the effectiveness of the weapon, how Oppenheimer was cheered by the entire staff of the laboratory like a 'prize fighter', and how they enjoyed 'a hastily arranged champagne dinner'. 'The only reaction I remember,' Richard Feynman recalls, '. . . was a very considerable elation and excitement. . . . I was involved in this happy thing, drinking and drunk, sitting on the bonnet of a jeep and playing drums, excitement running all over Los Alamos at the same time as the people were dying and struggling in Hiroshima' (Easlea, 1983, p. 112). The principal reason for the establishment of the Manhattan Project was the fear that Nazi Germany would develop atomic weapons. Easlea makes much of the fact that work on the atomic bomb actually intensified after Nazi Germany had surrendered to the Allied armies.

For Easlea, this behaviour can be accounted for in terms of these male scientists substituting for their lack of feminine procreative power, that is, 'womb envy'. Men 'give birth' to science and weapons to compensate for their lack of the 'magical power' of giving birth to babies. He argues that this is demonstrated by their pervasive use of aggressive sexual and birth metaphors to describe their work, such that the first uranium bomb, which was dropped on Hiroshima, was named 'Little Boy'. This imagery signifies unconscious male motivations, 'phallic psychology', which makes such technical inventions possible. Men are obsessed with gaining power and glory and he approvingly quotes Simone de Beauvoir's view that male accomplishments in the field of science and technology serve to bestow a virile status on the respective male achievers and thereby underwrite a claim to masculinity.

Unfortunately Easlea's account is strongly suggestive rather than analytical. As Adam Farrar (1985, p. 61) comments, these birth and rape metaphors 'only show that the means of representing significant practices in a male dominated culture are constructed in terms which are significant to men. They don't show that the practices so represented are necessarily masculine'. Although thought-provoking, Easlea's examination of the arresting metaphors used by scientists when describing their experiences of doing creative work does not, by itself, constitute grounds for arguing that male creativity is inherently about rape and violence. Individual motivations cannot be read directly from the rhetoric of science any more than the invention of the bomb can simply be explained in terms of the psycho-sexual anxieties of its particular male inventors.

Sexual imagery has always been part of the world of warfare and both the military itself and arms manufacturers are constantly exploiting the phallic imagery and promise of sexual domination that their weapons so conveniently suggest. This imagery however does not originate in particular individuals but in a broader cultural context. Easlea's analysis misses the social processes that give rise to this kind of masculinity and validate such scientific and technological projects in the first place. As Ludi Jordanova (1987, p. 156) comments, '[t]he interesting questions are how and why creativity of all kinds has been defined in a gender-specific way, and what implications this has for power relations'. That the technological enterprise has developed as a distinctly masculine realm may be largely a reflection of the male domination of all powerful public institutions, rather than something specific to the male spirit.

The language used by defence intellectuals when discussing nuclear strategy is particularly revealing. Carol Cohn (1987) discovered this recently when she spent a year in the company of defence strategists. Like Easlea she found the male world of nuclear planning suffused with sexual and patriarchal imagery and sanitized abstraction; a language designed to talk exclusively about weapons and not about human death. However, for her 'the interesting issue is not so much the imagery's psychodynamic origins, as how it functions' (Cohn, 1987, p. 695).She argues that this 'technostrategic' discourse serves to reduce anxiety about nuclear war by providing a series of culturally grounded and culturally acceptable mechanisms that distance the user from thinking of oneself as a victim, making it possible to think about the unthinkable.

> Language that is abstract, sanitized, full of euphemisms; language that is sexy and fun to use; paradigms whose referent is weapons; imagery that domesticates and deflates the forces of mass destruction; imagery that reverses sentient and nonsentient matter, that conflates birth and death, destruction and creation – all of these are part of what makes it possible to be radically removed from the reality of what one is talking about and from the realities one is creating through discourse. (Cohn, 1987, p. 715)

She correctly points out that it may be an illusion to assume that technostrategic language literally articulates rather than hides the actual reasons for the development and deployment of nuclear weapons. Rather than informing and shaping decisions, the discourse more often functions as a legitimation for political outcomes that have occurred for utterly different reasons. This language of patriarchal euphemisms permeates many spheres of high technology.

The Obsession with Control

In fact, there are many parallels between the ethos of the scientific community at Los Alamos and that of the computing fraternity. This is strongly reflected in Tracy Kidder's (1982) account of a group of men inventing a new computer in *The Soul of a New Machine.* Here again we find the mixture of professional competitive rivalry and complete dedication in the engineers' pursuit of the 'perfect computer' and in doing so, winning the race. Again it is a world of men working compulsively into the small hours, enjoying being stretched to the limits of their capacity, where there is no space for or compromise with life outside of work. It was 'the sexy job' to be a builder of new computers, and you had to be tough and fast; members of the group often talked of doing things 'quick and dirty', and of 'wars', 'shoot-outs', 'hired guns', and people who 'shot from the hip'. Sexual metaphors abound such that the excitement of working on the latest computer was likened to 'somebody told those guys that they would have seventy-two hours with the girl of their dreams'. It is surely no coincidence that the protagonists of the story are almost exclusively male.

It is evident that men identify with technology and through their identification with technology men form bonds with one another. Women rarely appear in these stories, except as wives at home providing the backdrop against which the men freely pursue their great projects. This masculine workplace culture of passionate virtuosity is typified by the hacker-style work so well described by Sherry Turkle (1984) in a chapter entitled 'Loving the Machine for Itself'. Based on ethnographic research at MIT, Turkle describes the world of computer hackers as the epitome of this male culture of 'mastery, individualism, nonsensuality'. 'Though hackers would deny that theirs is a macho culture, the preoccupation with winning and of subjecting oneself to increasingly violent tests make their world peculiarly male in spirit, peculiarly unfriendly to women.' (1984, p. 216). Being in an intimate relationship with the computer is also a substitute for, and refuge from, the much more uncertain and complex relationships that characterize social life. According to Turkle, these young men have an intense need to master things; their addiction is not to computer programming but to playing with the issue of control. It is about exerting power and domination within the unambiguous world of machinery.

Perhaps the ultimate illustration of men pitting themselves against machines is the story of the Air Force test pilots who became the first astronauts. What is absorbing about Tom Wolfe's account in *The Right Stuff* (1980) is its focus on the psychology of the test pilot. In flying aircraft higher and faster than they were designed to be flown, these men were constantly testing the limits of the physically possible, pushing 'the outer edge of the envelope' until the limits of the technology were reached and only having 'the right stuff' could save a man's life. This is the stuff that lets you function as a superhuman when you have pushed yourself beyond the edge of human and technical possibility, the stuff that allows you to feel in control in situations that were set up in advance as situations where control would be lost. And when they left the airfield these test pilots would push the outer edge in the male rituals of drinking and driving their cars at speeds almost out of control. The chances of dying were astonishingly high and yet these men actually were delighted to take on such odds and prove their courage.

One of the interesting twists to the story is the contrast between the description of Chuck Yeager, who was the first man to fly through the sound barrier and the trips of the first astronauts. Whereas Yeager ranked foremost among the true brothers of the right stuff, an astronaut was seen not as a pilot but a passive occupant of a rocket – shown by the fact that the first flight in a Mercury rocket was in fact taken by a chimpanzee. Wolfe tells of an incident when the seven astronauts insisted that a window be designed into the capsule which they could open themselves. They even demanded manual controls for the rocket so that they would have control and function more like pilots. In this sense, the commonly-drawn analogy between rockets and missile and the phallus is somewhat inappropriate. For all that the first astronauts were hailed as conquering heroes, according to their own codes of masculinity they had actually become subordinate to elaborate computerized controls, and were denied the scope to demonstrate their manliness. Real men fly planes, they don't just push buttons!

Forms of Masculinity

Although the above descriptions are all of relatively powerful groups of men developing or controlling key forms of technology, they actually involve a wide variety of workplace cultures and practices. What this points to is the need to distinguish between different forms of masculinity in relation to different areas of technology. To say that control over technology is a core element of masculinity is not to imply that there is one masculinity or one technology. There are diverse cultural expressions of masculinity just as there are diverse technologies. Masculinity, like femininity, takes historically and culturally specific forms. One need not presume that there is a single uniform behaviour pattern in all men to argue that the culture of technology is masculine. These disparate versions of masculinity reflect class divisions, as well as ethnic and generational differences.

The study of the social construction of masculinity has become a strong theme in the sociology of gender. A major contributor in this area, Bob Connell (1985, 1987), distinguishes between the culturally dominant forms of masculinity or 'hegemonic masculinity' and 'subordinated' or 'marginalized' forms. By 'hegemonic' he means a social ascendancy achieved not by force but by the organization of private life and cultural processes. It is the dominant cultural ideal of masculinity, which need not correspond closely with the actual personalities of the vast majority of men. Although Connell uses the term 'hegemonic masculinity' in the singular, I read him to suggest that there is a core of dominant masculinity which is reflected in different variants. In contemporary Western society, hegemonic masculinity is strongly associated with aggressiveness and the capacity for violence.

Of particular interest here is the extent to which control of technology is involved in this archetype of hegemonic masculinity. The cult of masculinity which is based on physical toughness and mechanical skills is particularly strong in the shop-floor culture of working-class men. All the things that are associated with manual labour and machinery - dirt, noise, danger - are suffused with masculine qualities.[2] Machine-related skills and physical strength are fundamental measures of masculine status and self-esteem according to this model of hegemonic masculinity. But although machismo on the shop-floor is one form of the dominant ethos of masculinity, it is not the only important one and in the next section we will go on to consider the ideologies and practices of militarism, where weapons are central to the techniques of violence and are almost entirely in the hands of men.

Let us first return for a moment to the example of computer hackers and look more closely at the way manliness is represented here. One might initially describe their form of masculinity as the professionalized, calculative rationality of the technical specialist. What is interesting for our purposes is the way they mythologize their work activities in terms of the traditional 'warrior ethic' of heroic masculinity. The construction of the heroic is usually around matters of combat and violence between men. In fact, these mainly white middle-class men are nowhere near real physical danger yet they are drawing on the culturally dominant form of masculinity for their notions of risk, danger and virility to describe their work. An apparent paradox emerges however when we look more closely at these descriptions of hackers. Seen through the eyes of Turkle or Weizenbaum these elite technologists are pictured as unattractive and pathological: 'bright young men of disheveled appearance. . . . Their rumpled clothes, their unwashed and unshaven faces, and their uncombed hair all testify that they are oblivious to their bodies and to the world in which they move.' (Weizenbaum, 1976, p. 116) They are 'losers' and 'loners' whose immersion in the world of machines has cut them off from other people, and they rely on the hacker subculture for their sense of identity.

The question that this poses is whether for these men technical expertise is about the realization of power or their lack of it.[3] That in different ways both things are true points to the complex relationship between knowledge, power and technology. An obsession with technology may well be an attempt by men who are social failures to compensate for their lack of power. On the other hand, mastery over this technology does bestow some power on these men; in relation to other men and women who lack this expertise, in terms of the material rewards this skill brings, and even in terms of their popular portrayal as 'heroes' at the frontiers of technological progress. By providing a largely psychoanalytical account of hackers, Turkle's notion of failure is very individualistic and does not address the wider cultural context within which hackers operate.[4] In particular, there is little mention of the extent to which race, class, and age matter in what counts as failure for men.

In our culture, to be in command of the very latest technology signifies being involved in directing the future and so it is a highly valued and mythologized activity. The mastery of other kinds of technology, such as that often found amongst working-class lads who are adept with cars, does not convey the same status or agency. Neither in fact does hegemonic masculinity, which is more strongly possessed by working-class than ruling-class men. The exaggerated masculinity found amongst working-class cultures must be viewed against the background of their relative deprivation, their low status and their comparative powerlessness in the broader society. The point here is that although technical expertise is a key source of power amongst men, it does not override other sources of power, such as position in the class structure.

There is one final but crucial point to be made about these masculine cultures of technology, which is that the ideology of masculinity is remarkably flexible.[5] A good illustration of this point is provided by engineering. Of all the major professions, engineering contains the smallest proportion of females and projects a heavily masculine image hostile to women.[6] Engineering is a particularly intriguing example of an archetypically masculine culture because it cuts across the boundaries between physical and intellectual work and yet maintains strong elements of mind/body dualism.

Central to the social construction of the engineer is the polarity between science and sensuality, the hard and the soft, things and people. This social construction draws on the wider system of symbols and metaphors which identify women with nature and men with culture. Sexual ideologies and stereotypes are diverse and fluid, but such opposites as 'male/female' and 'reason/emotion' are central to Western culture. The notion that women are closer to nature than men contains various elements such as that women are more emotional, less analytical and weaker than men. In the advanced industrial world, where scientific and technical rationality are highly valued, these associations play a powerful role in the ideological construction of women as inferior.[7] Like Turkle, Sally Hacker (1981) found that engineers attributed values in the social hierarchy on a continuum, giving most prestige to scientific abstraction and technical competence and least to feminine properties of nurturance, sensuality and the body. Engineering seems to be the very epitome of cool reason, the antithesis of feeling. Yet, as Hacker (1989) remarks, and as we have already seen, 'technics can be exhilarating, a source of intense pleasure, even arousal, at the core of innovation'. In *The Existential Pleasures of Engineering*, Samuel Florman (1976) stresses the sensual and physical, as well as intellectual, pleasures derived from practising engineering.

Machines can clearly evoke powerful emotions and sensual delight for men. One wonders if this is at least in part why ships and machinery are referred to by the female pronoun! Similarly, the complementary values of hard/soft are also used to legitimate female exclusion from the world of engineering.[8] Masculinity is expressed both in terms of muscular physical strength and aggression, and in terms of analytical power. 'At one moment, in order to fortify their identification with physical engineering, men dismiss the intellectual world as "soft". At the next moment, however, they need to appropriate sedentary, intellectual engineering work for masculinity too.' (Cockburn, 1985, p. 190)

No matter how masculinity is defined according to this ever-adaptable ideology, it always constructs women as ill-suited to technological pursuits.

Combat, the Heroic and Masculinity

We have been considering examples of the sense of mastery over technology and its connection with masculinity. If there is one institution in society that underwrites the ideology of hegemonic masculinity, it is the military. Contemporary Western society is suffused with popular images of men as enthusiastic killers – in Rambo-style fashions, war films, military toys and magazines. Weapons and particularly guns are the epitome of mastery as a means of domination. Guns are intrinsically associated with death and danger. Here the sense of mastery is enhanced by the closeness to physical danger, it is seen as the pinnacle of manly daring. The willingness of men to die – for their country, for their womenfolk, for their honour, is central to both military and manly values. War provides the ultimate test of manliness and is the legitimate expression of male violence. In this sense, the armed forces represent and defend the masculine ethic. Warrior values as expressed in armed combat are central to military mythology, which in turn is imbued with gender ideology.[9] Both war and weaponry are seen as exclusively male concerns, and the imagery surrounding them simultaneously portrays men as the brave warriors and women as the helpless wives–mothers–daughters, whose lives and honour the soldiers are fighting to protect.

Ideas about femininity and the unsuitability of women for the military are almost universal. At the core of all arguments against allowing women to perform combat roles are beliefs about the inferiority of the female body. Now that women are increasingly being recruited into the military, this search for sexual difference has, if anything, intensified. 'Western armed forces now conduct official studies of pregnancy, menstruation, and "upper body strength" in an

6.1 *Source:* Sophie Grillet

almost desperate search for some fundamental, intrinsic (i.e. not open to political debate) difference between male and female soldiers.' (Enloe, 1983, p. 138) According to military logic, combat involves lifting and carrying heavy things and pulling oneself over formidable obstacles. In a mirror image of non-military workplace ideology, assumptions about women's physical weakness and higher absenteeism are mobilized to exclude women from combat. As Cynthia Enloe comments, the distinctions between 'combat' and 'non-combat', the 'front' and the 'rear', which ultimately justify the military's sexual division of labour, are increasingly difficult to sustain in the face of modern warfare.

It was never easy however. Enloe (1983, p. 123) provides a wonderful illustration of the delicate manoeuvres that have been performed to maintain a male definition of combat. During the Second World War several anti-aircraft batteries were set up in Britain comprising men and women. The idea was that women could operate the guns' fire control instruments and so 'free' male soldiers to actually fire the guns. Thus, in the new mixed artillery crews women were assigned to fire control, searchlight operations, targeting and hit confirmation. These artillery women were defined as 'non-combat' personnel while the men standing next to them, but assigned to firing the guns, were designated as 'combat' personnel!

The view that women are constitutionally predisposed towards peace does not only feature in anti-feminist arguments. While distancing themselves from the ideology that defines women as physically inferior to men, many feminists also see war and soldiering as male and believe that women are naturally inclined to pacifism. It was noted at the beginning of this chapter that much feminist peace writing conceives of war and weaponry as the direct result of men's violent nature. Women, on the other hand, are seen as nurturing, cooperative and non-violent. Their role as mothers is supposed to lead women to value growth and preservation, as against death and destruction. Some writers are influenced by Carol Gilligan's (1982) view that women's cognitive and moral development is distinctly different from that of men. According to Gilligan, women's concrete and contextual style of thinking and moral reasoning involves an orientation toward nurturance and care, through relationship and connection. Thus appeals are made to women's 'caring' morality, or 'maternal thinking' in Sara Ruddick's (1983) version, to end war.

In constructing women as inherently peace loving, these feminists are implicity reinforcing a traditional model of masculinity and femininity. As authors like Genevieve Lloyd (1986) and Lynne Segal (1987) have pointed out, it is rather ironic that those ideas about psychological sex differences that are central to patriarchal ideology have now become so prevalent in popular feminist thinking.

> Femininity, as we now have it, has been constituted within the Western intellectual tradition to be what is left behind by ideals of masculinity, citizenship and patriotism. But if that is so the idea of a special antipathy between women and war has to be seen as in some ways a product of the very tradition to which it may now seem to be a reaction. It may be salutary to realise that the idea of the feminine that figures in some of the rhetoric of feminist peace groups springs from the same sources as General Barrow's conviction that there is no place for women on the battlefield. (Lloyd, 1986, p. 75)

As Jean Elshtain (1987) indicates, tales of women warriors and fighters are easily buried by the dominant narrative of bellicose men and pacific women. There are historical accounts of women military leaders who have actively led soldiers on the battlefield as well as contemporary examples of women combatants in national liberation armies around the world. As non-combatants too, most women have readily supported war aims and efforts, as they did during the two world wars. Although men are responsible for most war and violence, they are also responsible for most organized opposition to war. Many men, such as the conscientious objectors and male pacifists of the world wars, have rejected war and the military. Despite these countervailing tendencies, the received cultural images of women, men and war endure.

Indeed the recognition that the maleness of the military is only a façade throws this masculine ideology into sharp relief. The illusion is sustained by the refusal to acknowledge certain awkward facts. Despite the absence of women from the front lines, armed forces importantly depend upon women, both directly and indirectly as clerical workers, nurses, domestic and sexual services to male soldiers. During the two world wars women carried out technically skilled munitions work and even redesigned the weaponry they were making. Today women are again playing an important part in weapons production. Yet this dependence on women is systematically denied in our dominant cultural understanding of militarism. In fact, as weapons become more and more heavily based on electronics, the role of traditional male military 'virtues' is diminished - no enemy is ever seen, much less physically confronted - while simultaneously the salience of women's labour becomes ever greater. For the electronics industry is largely a women's industry, at least as far as production is concerned. Enloe (1983, p. 195) gives the example of a modern navy cruiser. Forty per cent of the cost of such a vessel is accounted for by its electronics, and without those electronics it would be useless in modern naval warfare. Thus, an archetypally male artefact is in reality built in large part by women.[10] A durable ideology of masculinity conceals the fact that although designed and controlled by men, military technology is increasingly produced by women. For as the modern battlefield becomes transformed by the destructive force of highly automated technology, the expression of masculinity as physical strength and aggression is increasingly overshadowed.

As we have seen repeatedly, technology is more than a set of physical objects or artefacts. It also fundamentally embodies a culture or set of social relations made up of certain sorts of knowledge, beliefs, desires and practices. Treating technology as a culture has enabled us to see the way in which technology is expressive of masculinity and how, in turn, men characteristically view themselves in relation to these machines. I have also described how, in order to maintain this male dominance over new and unfamiliar kinds of machinery, men willingly adapt and modify their ideas about masculinity. I will now explore women's relation to, or more specifically their absence from, this technical culture.

Women and Machines: Cognition or Culture?

Given the resilience of this association between technology and manliness, how do women think about and experience technology? What are the mechanisms, both formal and informal, that foster and reproduce the cultural stereotype of women as technologically incapable or indeed invisible in technical spheres?

The continuing male monopoly of weapons and mechanical tools is perhaps not so difficult to understand given the weight of male tradition and custom borne by these instruments of war and production. The old story that you had to be strong to work with machines had at least some credibility in this context.[11] The male dominance of new technologies is, at first sight, much more puzzling. It was a commonly held expectation that with the development of microelectronics, and the decreasing importance of heavy industrial technology, the gender stereotyping of technology would diminish.

Computing is a crucial example, because as a completely new type of technology it had the potential to break the mould. In terms of sexual divisions, there are three distinct paths along which this technology might have developed. Computing could have been gender-neutral with no basic differentiation between female and male users. Or, it could have been a technology that women appropriated. After all, the image of new electronic computer technology fits with femininity in that it is clean, sedentary work involving rote tasks, detail, precision and nimble typing fingers. Yet recent evidence on the gender gap in access to computers at school, at play and at home, supports the idea that our culture has already defined computers as preeminently male machines. Numerous British, American and Australian surveys show that boys vastly outnumber girls wherever there is discretionary use of these machines such as in school computer clubs, computer summer camps, at home and in games arcades.[12] In response to this disturbing trend, a number of feminist researchers have recently investigated the relation of women and girls to computers. These studies afford useful insights into the marginalization of women from technology more generally.

Computing Inequality at School

Although the work-related cultures which we have examined have their own dynamics, they are also the result of cultural processes that take place outside of work and that are carried into it. Technologies, like people, are already sex-typed when they enter the workplace. Most women never approach the foreign territory of these masculine areas. This sex segregration at work reflects the fact that patriarchal relations are an integral part of our entire social system. In modern societies it is the education system, in conjunction with other social institutions, which helps to perpetuate gender inequalities from generation to generation. Schooling, youth cultures, the family and the mass media all transmit meanings and values that identify masculinity with machines and technological competence. These social contexts are intertwined and mutually reinforcing, but they should not be seen simply as external forces. Individuals actively participate in, resist, and even help reproduce by resisting, these social practices. An excellent illustration of the relationship between sexuality, schooling, youth cultures and employment is provided by Paul Willis in *Learning to Labour* (1977). He describes how working-class boys end up in working-class jobs by stressing the role that the demonstration of masculinity plays in valorizing low status labour. The class and gendered nature of traditional manual labour is created, maintained and reproduced through these interconnected cultural processes.

How does a completely new technology like computers, that may not automatically conform to preexisting patterns of gender differentiation, fit into these processes? Focusing on computers may enable us to see more clearly the social mechanisms through which a new technology becomes integrated into the masculine cultural system.

Schools are the most obvious places where young people first come in contact with computers. There is now an extensive literature on sex stereotyping in general in schools, particularly on the processes by which girls and boys are channelled into different subjects in secondary and tertiary education, and the link between education and gender divisions in the labour market. As with scientific and technological areas of tertiary education generally, the sex ratio of computer science is very marked. Since at least the mid-1970s there have been anti-discrimination legislation, equal opportunity programmes and other government and non-government initiatives in many countries to redress this imbalance. Despite all this effort, the number of girls taking computer science at British universities has been decreasing. The proportion of female applicants for undergraduate level computer science courses dropped from approximately 28 per cent in 1978 to 13 per cent in 1986 (Hoyles, 1988, p. 9). In fact there are fewer women applying for computer science courses now than nine years ago although the subject has doubled in size. Similar situations are found in other countries.

This drop appears to be linked to the widespread introduction of microcomputers into schools. Here girls quickly learn that computers are 'just for the boys'. Numerous investigations into the under-representation of girls in science have indicated how the presentation of the subject alienates girls.[13] Computers have been linked to things scientific and mathematical, traditionally male subjects. There has been a tendency to site school microcomputers in the science/maths department and computer studies is almost always taught by mathematics teachers, usually male. Even though it is now generally recognized that ability in mathematics is not an indication of aptitude for computing, it is still taken into account for entry to computing courses at school. 'Thus computers tend to be conceptually assimilated to the category of science, mathematics and technology and acquire some of the traditional qualities of differentiated interest amongst boys and girls' (Hoyles, 1988, p. 10).

Gender differences in educational experience are not simply the result of what is taught in courses of formal instruction. In a more profound way the culture of the school is involved in constructing gender and sexuality through the 'hidden curriculum' – teaching in an implicit way meanings and behaviours associated with femaleness and maleness, with femininity and masculinity. Studies of classrooms show that teachers behave differently to girls and boys, they speak to them differently, they require different responses and different behaviour from them. Gender identity is profoundly important to children's perception of themselves. Girls feel the need to display a set of behavioural patterns that are perceived as being feminine: these feminine qualities are however incompatible with the qualities supposed necessary for a 'mathematical mind'.[14] Girls internalize the belief that boys possess something that they lack; difference is lived as inferiority. One study of primary school children reported, for example, that '[g]irls actually believed that boys were naturally ordained with a profusion of masculine esoteric skills such as being able to drive a car, tractor or helicopter' (Clarricoates, 1980, p. 39). Computers are seen as belonging to the realm of machinery and mathematics – a daunting combination for girls.

However there is a danger here of implying that, in conforming to the gender stereotype and thus rejecting technology, girls are their own worst enemy. Feminists have now challenged this passive model of female socialization, arguing that girls may well use their femininity as a form of resistance at school, or even resist feminine roles themselves (Barton and Walker, 1983). Some girls are interested in computers but it is difficult for them to pursue this because boys actively and aggressively capture computer time where, as is usually the case, there is insufficient computer supply in schools. This harassment of girls interested in computing continues into tertiary education. At this stage the harassment takes the form of obscene computer mail or print-outs of nude women. Women students in computer science at MIT found this problem so pervasive that they organized a special committee to deal with it.

'Space Wars' – Games for the Boys

Children today are more likely to develop their interest in information technology at home than at school. Schools reinforce the early socialization into gender roles that takes place within the family.

Many children's toys encourage boys to be assertive and independent, to solve problems, experiment with construction and, more recently, to regard the technological aspects of their toys with confidence and familiarity. The skills which children learn from these toys lay the foundations of mathematical, scientific and technological learning. By contrast, 'girls' toys' such as dolls foster different skills which are associated with caring and social interaction. Just as boys often come to school with the advantage of having played with mechanical toys, or connected an electric train set, they now have often played video games at home. Toys are an important part of the differentiated learning experiences between girls and boys. These toys in turn reflect the division of labour between women and men within the family. In the chapter on domestic technology we saw the way in which household technologies are sharply gendered. Technologies of external household and car maintenance are traditionally the husbands' sphere, while women primarily use the technologies of the kitchen and cleaning. Moreover, control of technologies of entertainment such as the television and video recorder are also gendered male.

Computers all too easily fit in to this sex-stereotyped view of technology. There is a tendency for the home micro to be bought for the sons of the family. This is encouraged by advertisements for computer games and home computers which are aimed at a male market and often feature pictures of boys looking raptly at the screen. Evidence collected by the Equal Opportunities Commission in 1985 revealed that of all British households owning micro computers, boys are 13 times more likely than girls to be using them. Moreover, only 4 per cent of micros are used by their mothers. Children quickly learn from their parents which are the appropriate spheres for them. In a California survey in which school children were asked to describe how they would use computers when they were 30 years old, 'the boys said they would use them for finances, data processing, and games; the girls thought they would use them for housework. Wrote one sixth grade girl: "When I am thirty, I'll have a computer that has long arms and that can clean the house and cook meals, and another to pay for groceries and stuff." ' (Kolata, 1984, p. 25)

Games are the primary attraction of computers for children. Given that it is men (often computer hackers) who design video games and software, it is hardly surprising that their designs typically appeal to male fantasies. In fact video games began at one of the places where computer culture itself got started. The first video game was Space War, built at MIT in the early 1960s.[15] Many of the most popular games today are simply programmed versions of traditionally male non-computer games, involving shooting, blowing up, speeding, or zapping in some way or another. They often have militaristic titles such as 'Destroy All Subs' and 'Space Wars' highlighting their themes of adventure and violence. No wonder then that these games often frustrate or bore the non-macho players exposed to them. As a result, macho males often have a positive first experience with the computer; other males and most females have a negative initial experience.

It is this masculine narrative content of much computer game software that has received the most attention in explanations of the difference between female and male interest in video games. As we shall see below, many analyses focus on the private experience or 'intimate relation' with the machine ignoring the social dimension of interest in computing or in playing games. The predominantly male interest in games is a function of time and a legacy of male adolescent culture. Overall, girls simply have fewer opportunities to use computers than boys because the experience of leisure time is deeply divided along sex boundaries. Like their mothers, girls have a lot less time to play at home because of their domestic responsibilities. Young working-class daughters are expected to help with childcare and other household tasks in a way that their brothers are not. Boys learn from their fathers that it is their right to concentrate totally on the computer if they choose, oblivious of the surrounding domestic environment. Males are more easily allowed to follow up interests which do not have to be justified as benefiting anyone else.

In addition, girls' extracurricular activities are generally much more restricted than boys. Parents are cautious about allowing girls to stay after school in the unstructured environment of computer clubs.[16] Public places like video arcades, which are central to the leisure culture of young male adolescents, are virtually off limits for most girls. They are populated almost exclusively by males; the few females in evidence are usually spectators. Leslie Haddon has shown that it was the continuity from pinball machines that helped shape the arcade game-playing world as a predominantly masculine one. Electronic games directly appropriated the role of pinball and, within a few years of their introduction, pinball sales had declined by two-thirds. The institutions that these young males had built up around pinball - the values, rules, and rituals - were transferred to the video game. '[T]he location of video games within the arcade and certain other contexts had meant that the new machines were incorporated into the existing social activities of this milieu. Amusement parks, and many of the other public sites where coin-operation machines were found, were part of street culture. They were mainly male, particularly young male, preserves.'[17] Thus the new technology was slotted into a pre-existing male subculture and took on its masculine face.

Mastery of the Machine: Vive la Différence?

Throughout this chapter I have been arguing that cultural factors are important in understanding the masculinity of technology. By this I mean that the absence of technical confidence or competence does indeed become part of feminine gender identity, as well as being a sexual stereotype. Using the instructive example of computers, I have explored the interrelated social processes that make this technology into an alien culture for girls. It is now time to consider an argument that has been enthusiastically received by many Western feminists - that technical performance is a feature of fundamental cognitive difference between the sexes.

There have been endless variations on the theme that men's superior achievement demonstrates their greater physical and mental capacities. Traditionally, the significant discrepancy between the sexes in their ability to work with technology was attributed to physical strength or weakness and feminists spent the best part of the 1970s discrediting this doctrine of natural difference. I am prompted to wonder if it is merely an accident of history that, just as there is a major shift in the nature of technology from industrial to information technology, an increasing number of feminist accounts of women and computers are themselves emphasizing cognitive sex-differences. These alleged differences between the sexes are conceptualized as opposed pairs which connect with other sets of oppositions. Males are portrayed as fascinated with the machine itself, being 'hard masters' in terms of computer programming, followers of rules and competitive. Females are described as only interested in computers as tools for use and application, as 'soft masters', as more concrete and cooperative in orientation.[18] Thus it is argued that girls are less likely to achieve, not simply as the result of biological difference, but because of essential psychological differences. These arguments are reminiscent of two views that are by now somewhat discredited. One is the old sexual stereotype about women being too emotional, irrational and illogical, not to mention lacking the visual spatial awareness, to be good at mathematics: the other is the 1960s and 1970s belief that working-class and black children were naturally suited to less abstract or more concrete forms of learning. The new and fundamentally feminist twist in the argument, as we shall see, is that difference is no longer equated with inferiority or hierarchical ordering.

By far the best exposition of this view, and one which is drawn on widely by other authors, is to be found in the work of Sherry Turkle (1984, also Turkle and Papert, forthcoming).[19] From her observations of young children programming at school, Turkle found that boys and girls tended to use two distinctive styles of computing, which she calls 'hard' and 'soft' mastery. Hard masters are overwhelmingly boys, imposing their will over the machine by implementing a structured, linear plan. The goal is to control the machine. Girls tend to be soft masters, having a more 'interactive', 'negotiating' or 'relational' style. They relate to the computer's formal system as a language for communication rather than as a set of rigid rules. Turkle draws a parallel with Claude Lévi-Strauss' distinction between Western science and the science of preliterate societies in terms of the contrast between planning and *bricolage* or tinkering. 'The former is the science of the abstract, the latter is a science of the concrete. Like the *bricoleur*, the soft master works with a set of concrete elements. While the hard master thinks in terms of global abstractions, the soft master works on a problem by arranging and rearranging these elements, working through new combinations.' (Turkle, 1984, p. 103) Turkle is clear and emphatic that neither of these styles is superior for programming – they are different, and diversity or 'epistemological pluralism' should be celebrated. The problem for women then is the differential value accorded to the different styles. Computer expertise is defined as hard mastery; it is recognized as the only correct way to programme. Soft mastery is culturally constructed as inferior. Once more women are not up to hard male mastery.

Turkle is correct to point out that when gendered styles of computing are identified by teachers, they are valued accordingly. When women and girls do have a facility with programming, the categories for evaluating their performance are themselves gender-biased. They are designated as getting the right results by the wrong method. Only male mastery is identified as the rational, logical approach. However I am uneasy when Turkle argues that male 'planners' versus female 'tinkerers' represent basic cognitive styles that are grounded in psychological sex differences. Here she is influenced by the work of Evelyn Fox Keller (1983, 1985), Nancy Chodorow (1978) and perhaps most by Carol Gilligan (1982). In different ways these authors all pose an essential theory of sexual difference in cognitive skills.

To the extent that this signals 'the return to conventional ideas of fundamental and comprehensive cognitive, emotional and moral difference between women and men' (Segal, 1987, p. 146) I am unconvinced. Firstly, on purely empirical grounds I am sceptical about the evidence provided for the existence of sex differences in cognitive styles. For example, Martin Hughes et al. (1988) found no such differences. Previous research on sex differences with regard to mathematical ability has always stressed the lack of confidence and the conformity of girls and the resulting tendency for them to follow the rules diligently. This would lead one to suppose that girls would be the hard masters in computing. Are we now to believe that in computing boys follow the rules and girls are practising an alternative style?

More generally the search for 'significant' sex differences in this or that behaviour has a doubtful political pedigree and it is difficult to avoid the conclusion that such research finds what it has set out to find. Although studies do find evidence of differences between the sexes, the variation within the sexes is more important than the differences between them.[20] Secondly, and more fundamentally, it is increasingly clear that cognition cannot be stripped of its social content to reveal pure logical reasoning.

Over the last ten years or so developmental psychology has recognized that the development of children cannot be understood outside the social context in which it occurs.[21] Social relationships, understandings and practices play a constitutive role in the elaboration of the child's conceptual knowledge. To present differences in programming style as differences of individual psychology, as Turkle does, is to assume an individualized account of learning (Linn, 1985, p. 95). Learning is a collective, social process. Turkle's predominantly psychological rather than sociological framework leads her to neglect the historical and cultural context in which computing education takes place. The pattern of boys being more independent and strategy-oriented, and girls being more concrete and dependent, bears a striking resemblance to the differences discussed by Valerie Walkerdine (1989) in the cognitive styles expected of, and encouraged in, boys and girls by teachers in the primary school. In our society the computer has become socially constructed as a male domain; children learn from an early age to associate computers with boys and men. This means that girls approach the computer less often and with less confidence than boys. It may also mean that there are significant gender differences in how girls and boys relate to the machine and what it means to them. They may even have a tendency to want to use the machine for different things. But we should be extremely wary of saying that because women have different ways of proceeding, this indicates a fundamental difference in capacity. Rather, such discrepancies in cognitive style as can be observed are the consequence of major sexual inequalities in power.

In this connection it is salutary to note that the very first computer programmers were women. Between 1940, and 1950, many women were engaged in programming, coding, or working as machine operators. Again it was due to the exigencies of war that women were recruited by the military into both civilian and military positions to work as trained mathematicians to calculate firing tables by hand for rockets and artillery shells. When ENIAC (Electronic Numerical Integrator and Calculator), the first operational computer, was built in the United States in the early forties, these women were assigned to programme it and became known as the 'ENIAC girls' (Kraft, 1979, p. 141). It was because programming was initially viewed as tedious clerical work of low status that it was assigned to women. As the complex skills and value of programming were increasingly recognized, it came to be considered creative, intellectual and demanding 'men's work'. Thus, depending on the circumstances, different cognitive styles may be characterized as 'masculine' or 'feminine' according to the power and status that attaches.

Throughout this book I have been arguing that technology is more than a set of artefacts. Technology is also a cultural product which is historically constituted by certain sorts of knowledge and social practices as well as other forms of representation. Conceiving of technology as a culture reveals the extent to which an affinity with technology has been and is integral to the constitution of male gender identity. Masculinity and femininity are produced in relation to each other and what is masculine, according to the ideology of sexual difference, must be the negation of the feminine. Different childhood exposure to technology, the prevalence of different role models, different forms of schooling, and the extreme segregation of the labour market all lead to what Cockburn describes as 'the construction of men as strong, manually able and technologically endowed, and women as physically and technically incompetent' (1983, p. 203).

Gender is not just about difference but about power: this technical expertise is a source of men's actual or potential power over women. It is also an important part of women's experience of being less than, and dependent on, men. However, it should be remembered that the construction of masculinity is a complex process. There is not one monolithic masculinity and not all men are competent with technology. Rather, technical competence is central to the dominant cultural ideal of masculinity, and its absence a key feature of stereotyped femininity. The correspondence between men and machines is thus neither essential nor immutable, and therefore the potential exists for its transformation.

NOTES

1 For example, see Dorothy Dinnerstein's (1976) analysis of men's war-making activities in terms of their desire to appropriate from women the power of giving life and death.
2 Willis (1977) and Cockburn (1983) contain excellent discussions of masculinity in the industrial workplace.
3 Maureen McNeil (1987, p. 194) makes this point in a review of Cynthia Cockburn's (1985) book.
4 See Pam Linn's (1985) interesting article on microcomputers in education which makes these points in relation to Turkle's study.
5 See chapter 3 where I discussed the historical variability of the gendering of jobs. See P. Kraft (1977)and J. Greenbaum (1979) on the history of computing which provides an excellent illustration of this. Over time, jobs that were firmly characterized as suitable for men become feminized and 'women's work' becomes men's.
6 Although this is the case in most Western countries, the proportion of women engineers in Eastern Europe is significantly higher.
7 It should be noted that there are also distinct national cultures of engineering. In a recent article, Eda Kranakis (1989) contrasts French and American attitudes towards manual labour throughout the nineteenth century. Whereas machine builders and mechanical engineers in France bore a certain social stigma, this was not the case in the United States. In fact, the locomotive driver, Casey Jones, was portrayed as a heroic, romantic figure in American folk culture. Although Kranakis does not touch on gender, such national differences must inform stereotypes of masculinity.

8 See also Game and Pringle's (1983, pp. 28–32) discussion of the polarities of heavy/light, dirty/clean, mobile/immobile which are broadly associated with masculinity and femininity.

9 It should be noted that the boys and men who are typically recruited or pressed into service as foot soldiers or ships' crews are drawn from the relatively powerless strata of societies. As Enloe points out, elite males who serve as officers try to use male camaraderie to reduce the all-too-obvious class and ethnic tensions among their troops. See Cynthia Enloe (1980).

10 In Silicon Valley the majority of women are black, Hispanic, or Asian. Furthermore, women electronics workers are also extensively employed in countries such as the Philippines, South Korea, Taiwan, Singapore and Indonesia. For details, see Enloe (1983, chapter 7).

11 Although to my amazement I recently heard a chess expert arguing that men were better than women at chess because of their superior 'visual-spatial' ability. How one would account for the success of computer chess according to this line of argument is beyond me!

12 See, for example, *Sex Roles*, Volume 13, Nos 3–4, August 1985, Special Issues: 'Women, Girls, and Computers'.

13 In Britain, Alison Kelly is well known for research in this area. See, for example, her edited collection (1981). The Girls and Mathematics Unit at the London Institute of Education has also done excellent work in this area; see for example, V. Walkerdine (1989). In this book they dismiss the commonly held view that boys are better than girls at mathematics. 'Girls, at the nexus of contradictory relationships between gender and intellectuality, struggle to achieve the femininity which is the target of teachers' pejorative evaluation. They often try to be nice, kind, helpful and attractive: precisely the characteristics that teachers publicly hold up as good. . . . while privately accusing the girls of doing precisely these things. Thus they are put in social and psychic double-binds. Few girls achieve both intellectual prowess and femininity.' (p. 203) Their central thesis is that girls' attainment is undervalued because the categories for evaluating performance and achievement are themselves gender-biased.

14 It should be noted that although we are talking about the development of femininity and masculinity as a general process, there is diversity within them, especially as they are modulated by class and racial divisions. This point is made in several of the articles in Rosemary Deem (1980) and Madeleine Arnot and Gaby Weiner (1987).

15 Interesting discussions of video games are contained in chapter 2 of Turkle (1984) and Haddon (1988).

16 This may be particularly important in Britain where computers entered the school curriculum later than in the USA. Often it was school computer clubs which were first set up and were dominated by boys. Girls were more likely to be introduced to computers afterwards in the classroom.

17 See Haddon (1988), p. 211. Haddon traces the roots of the home micro

back to both early hobbyist machines in the UK and USA and through the various lineages of interactive games. He particularly examines the rise of games-playing as the dominant application and argues that the marketing strategy of the early British producers was geared to the strong male hobbyist tradition.

18 For example, Deborah Brecher (1988) argues that there are gender-based differences in learning computing; boys' style of learning is rule-based; girls' style is holistic. At the conference on which the book is based there was frequent discussion of the ways women's more human-oriented tool approach would result in improved software design.

19 For two recent articles which draw on the work of both Turkle and Gilligan, see L. Lewis (1987) and Sutherland and Hoyles (1988).

20 For an excellent critique of sex difference research, including Gilligan's work, see C. Fuchs Epstein (1988), especially chapter 4.

21 The essays in Richards and Light (1986) share this perspective.

Conclusion

This book is intended as a contribution to both academic and political debates about the connections between gender, technology and society. Drawing on perspectives from radical science to radical feminism, I have argued that the use/abuse model that represents technology itself as neutral, and asserts that it is the human application of technology that determines whether it has beneficial or destructive effects, does not go far enough. By contrast, the social shaping approach insists that technology is always a form of *social* knowledge, practices and products. It is the result of conflicts and compromises, the outcomes of which depend primarily on the distribution of power and resources between different groups in society. Although there are other equally powerful forces shaping technology, such as militarism, capitalist profitability, and racism, this book has concentrated on gender. Nuclear weapons, for example, are the product of both the military–industrial complex and patriarchal culture.

The sociology of technology can only be strengthened by a feminist critique. This means looking at how the production and use of technology are shaped by male power and interests. It also means broadening the definition of technology, and tracing the origins and development of 'women's sphere' technology that have often been considered beneath notice. In common with most sociological research, the sociology of technology still suffers from a male bias that is largely interested in manufacturing and, more recently, military technology.

However, the search for a general feminist theory of technology, or of science, is misguided. It is important to show that the development of technology has been mediated by gender power relations, but the dangers of attempting a more general theory should be apparent from the preceding chapters. Instead, I have argued that we need to analyse the specific social interests that structure the knowledge and practice of particular kinds of technology. As more empirical work is done, it may be possible to draw further links between the ways men's interests influence different areas of technology. Such research is in its infancy and I hope this book encourages its growth.

Sociology's partial analysis of factors bearing on the development of technology has its corollary in an inadequate political response to technology. The development of a critique of the technological determinism implicit in much of both the sociological and feminist literature is thus politically apposite. For the notion that technology is a neutral force determining the nature of society is a depressing one, robbing us of any power to affect its direction. Rather than seeing technology as the key to progress or, more recently, the road to ecological or military destruction, the social shaping approach provides scope for human agency and political intervention.

Having said that, it may appear that the politics implicit in my account are profoundly pessimistic. For if technology is imprinted with patriarchal designs, what is to be done? To answer this I must firstly reiterate that the relationship between technological and social change is fundamentally indeterminate. The designers and promoters of a technology cannot completely predict or control its final uses. There are always unintended consequences and unanticipated possibilities. For example when, as a result of the organized movement of people with physical disabilities in the United States, buildings and pavements were redesigned to improve mobility, it was not envisaged that these reforms would help women manoeuvring prams around cities. It is important not to underestimate women's capacity to subvert the intended purposes of technology and turn it to their collective advantage. Although the telephone was developed and marketed to directly duplicate the functions of the telegraph, we have seen how women primarily use the telephone for sociability. Recognition of such contradictions and the space they create for change is particularly important in avoiding political pessimism.

Identifying the gendered character of technology need not lead to a rejection of existing 'patriarchal' technology. Neither does it require us to abstain from working 'in' technology. This is where I part company with those feminists who adopt an essentialist position that seeks to base a new technology on a fixed and universal set of women's values. For instance, eco-feminism maintains that women are closer than men to nature and that the technologies men have created are based on the domination of women as well as nature. This approach locks us into a double bind: technology is irredeemably masculinist, exploitative and militarist, yet women need and want technological skills and competence. An appeal to an idealized femininity is no way out of this dilemma; rather than simply going 'back to nature', we need to work from within and without to create another kind of culture.

I believe that there is room for an effective politics around gaining access to technological work and institutions. There are opportunities for disruption in the engine-rooms of technological production. The involvement of more women in scientific and technological work, in technology policy, education and so on, may bring significant advances in redesigning technology and constitute a challenge to the male culture of technology. Working in these spheres does not necessarily entail cooption into the world of patriarchal values and behaviour. As the proportion of women engineers grows, for example, the strong relationship between the culture of engineering and hegemonic masculinity will eventually be dismantled. Although new forms of femininity and new patterns of dominance may emerge, even small changes have a cumulative effect on gender relations more broadly understood.

This is not to deny that women pay a high price for venturing into such male-dominated territory. For many women the price is too high – requiring them to sacrifice major aspects of their gender identity. No equivalent sacrifice has been expected of men. Their identification with technology has been taken for granted, women's absence cast as women's problem. But women's problem is men. Men have to learn that technology is not 'theirs' and give up the privileges and power that go with this construction of masculinity. Ultimately this depends on transforming gender power relations which in turn requires changing the nature of work itself so that childcare and housework can be equally shared. Access politics alone cannot succeed because the institutions themselves are founded on gender inequality.

Strategies to increase women's participation are anyway limited by the extent to which power relations are inscribed in the technologies themselves. Women's reluctance 'to enter' cannot be seen as irrational given that so much technological development is devoted to warmongering and making a profit at the expense of human beings and the environment. Much of the research on electronics in this century has been sponsored by the military, especially in the United States.[1] Military exigencies and military support have been crucial in the development of 'civilian' technologies, such as the digital computer. Even apparently pure science is often funded for military purposes. For example, one of the major reasons for research on the earth's gravitational field is how to make nuclear missiles more accurate so that they can destroy opposing nuclear forces in a first strike.

This points to the need for a more radical critique of technology itself. Certain kinds of technology are inextricably linked to particular institutionalized patterns of power and authority. Despite the fact that I have been arguing against essentialist values, the radical feminist contribution to the debate has been important in politicizing science and technology. However, in so far as their emphasis is exclusively on patriarchal relations, they presume that if women were in control of technology they would be able to apply these technologies in beneficial ways. For all that women might design better products, redesigning at the level of the individual product is limited by wider social and economic structures.

An integral part of these wider structures, which has been subjected to critical analysis by the women's movement and the radical science movement, is the notion of technical expertise. By unmasking technology's supposed neutrality, the social shaping approach demystifies the layers of expert knowledge that are pivotal to the power of various professions. The feminist analysis reveals the extent to which expertise is monopolized by men. Men's appropriation of technology is central to their privileged position in paid production. For example, I have argued that technology has been and is the key to consolidating the power of the male medical profession. Technology plays a different role in different masculinities, but the power relation is still there. Certain kinds of work experience, in particular men's, is recognized as 'technical' and legitimated as expertise. Women's knowledge and skills have been traditionally undervalued. Contesting this involves re-examining the accepted definitions of expertise and challenging the sexual division of labour which sustains them. Technical competence is certainly not the only source of male power, but it is an important one, especially in relation to women.

Breaking the nexus between experts and technology is one of the projects of the 'alternative technology' movement. Attempts to design 'socially useful products' have opened up technology to wider public participation. In doing so they have given rise to a profound questioning of the nature of technology itself, the design methods used, and the way work is organized. Although this movement has many strands, its aim is to put social use and need before profit and involve working people in the processes of technological planning, design and production. Although these initiatives have lent substance to the possibility of developing different kinds of technology, they have concentrated on producing skill-enhancing technologies for male craft workers. As such, they have questioned neither the masculinity of skill definitions nor what counts as a product.[2] Women's interests and participation have been marginalized; a more democratic and egalitarian society would be reflected in their increased involvement in decision-making about technology at the workplace, at home and in the community.

Designing alternative feminist technologies is, however, far from straightforward. Just as the campaign for socially useful products in a capitalist context can only begin to specify the criteria by which to judge social use and need, so too our conceptions of a technology based on women's interests in a patriarchal society are necessarily embryonic. Feminine values are themselves distorted by the male-dominated structure of society. Rejecting essentialist notions of values as inherently masculine or feminine opens up debate about the form that values, such as caring and nurturing, should take. These forms will be different from existing forms of femininity, such as putting men's and children's needs first, that have relations of subordination built into them. Rather than calling for a technology based on feminine values, we need to go beyond masculinity and femininity to construct technology according to a completely different set of socially desirable values.

Feminist debates about political strategy concerning technology posit forms of action that break with conventional politics. They are about making interventions in every sphere of life. This means contesting the direction and use of technology around a whole range of particular locations - from the workplace to the health clinic, the school canteen to the supermarket, from the design of housing and vacuum cleaners to the design of sewerage systems. Small victories can make an enormous difference to people's experience and are politically achievable - such as changing the design of buses so that women with prams can travel more easily.

The time is ripe for reworking the relationship between technology and gender. The old masculinist ideology has been made increasingly untenable by the dramatic changes in technology, by the challenge of feminism and by the new awareness of the vulnerability of the natural world. I have argued that technologies reveal the societies that invent and use them, their notions of social status and distributive justice. In so far as technology currently reflects a man's world, the struggle to transform it demands a transformation of gender relations.

NOTES

1 Since World War II, as much as 40 per cent of research and development effort worldwide is devoted to the military. A recent calculation estimates that 37 per cent of the British engineering industry is reliant on military markets. See David Dickson (1984) for a description of the ever-closer relationship between science institutions and military–industrial interests in the USA.
2 See Pam Linn's (1987) instructive discussion of these schemes.

Bibliography

Ahmed, I. (ed.) 1985: *Technology and Rural Women*. London: Allen and Unwin.

Allan, G. and Crow, G. 1990: 'Constructing the Domestic Sphere: The emergence of the modern home in post-war Britain' in Corr, H. and Jamieson, L. (eds) *Politics of Everyday Life: Continuity and Change in Work and Family*. London: Macmillan.

Amram, F. 1984: 'The Innovative Woman'. *New Scientist*, 24 May 1984, pp. 10–12.

Andrews, W. and D. 1974: 'Technology and the housewife in nineteenth-century America'. *Women's Studies, 2*, pp. 309–28.

Arney, W. 1982: *Power and the Profession of Obstetrics*. Chicago: University of Chicago Press.

Arnold, E. and Burr, L. 1985: 'Housework and the Appliance of Science' in W. Faulkner, and E. Arnold (eds) 1985.

Arnot, M. and Weiner, G. (eds) 1987: *Gender and the Politics of Schooling*. London, Hutchinson.

Baran, B. 1987: 'The Technological Transformation of White-Collar Work: A Case Study of the Insurance Industry' in H. Hartmann et al. (eds) 1986, 1987 vol. 2.

Barnes, B. and Edge, D. (eds) 1982: *Science in Context: Readings in the Sociology of Science*. Milton Keynes: Open University Press.

Barton, L. and Walker, S. (eds) 1983: *Gender, Class and Education*. Lewes: Falmer Press.

Bebel, A. 1971: *Woman under Socialism*. New York: Schoken Books.

Bereano, P., Bose, C. and Arnold, E. 1985: 'Kitchen Technology and the Liberation of Women from Housework' in Faulkner, W. and Arnold, E. (eds) 1985.

Bijker, W., Hughes, T. and Pinch, T. (eds) 1987: *The Social Construction of Technological Systems*. Cambridge, Mass.: MIT Press.

Bittman, M. 1988: 'Service Provision, Women and the Future of the Household'. Unpublished paper.

Bose, C., Bereano, P. and Malloy, M. (1984): 'Household Technology and the Social Construction of Housework' *Technology and Culture*, 25, pp. 53–82.

Braverman, H. 1974: *Labor and Monopoly Capital: The Degradation of Work in the Twentieth Century*. New York: Monthly Review Press.

Brecher, D. 1988: 'Gender and learning: Do Women learn differently?' in K. Tijdens, M. Jennings, I. Wagner, and M Weggelaar (eds): *Women, Work and Computerization: Forming New Alliances*. Amsterdam: North Holland.

Bruland, T. 1982: 'Industrial Conflict as a Source of Technical Innovation: Three Cases'. *Economy and Society*, 11, pp. 91-121.

Burnett, J. 1978: *A Social History of Housing 1815-1970*. Newton Abbot: David & Charles.

Chambers, D. 1983: 'Symbolic equipment and the objects of leisure images'. *Leisure Studies*, 2, pp. 301-15.

Chodorow, N. 1978: *The Reproduction of Mothering: Psychoanalysis and the Sociology of Gender*. Berkeley, Calif.: University of California Press.

Cichocki, M. 1980: 'Women's Travel Patterns in a Suburban Development' in G. Wekerle, R. Peterson, and D. Morley (eds): *New Space for Women*. Boulder, Colorado: Westview Press.

Clarricoates, K. 1980: 'The importance of being Ernest . . . Emma . . . Tom . . . Jane. The perception and categorization of gender conformity and gender deviation in primary schools' in R. Deem (ed.) 1980.

Cockburn C. 1981: 'The Material of Male Power'. *Feminist Review*, 9, pp. 41-58.

Cockburn, C. 1983: *Brothers: Male Dominance and Technological Change* London: Pluto Press.

Cockburn, C. 1985: *Machinery of Dominance: Women, Men and Technical Know-How*. London: Pluto Press.

Cohn, C. 1987: 'Sex and Death in the Rational World of Defense Intellectuals'. *Signs*, 12, 4, pp. 687-718.

Coleman, A. 1985: *Utopia on Trial: Vision and Reality in Planned Housing*. London: Hilary Shipman.

Connell R. 1985: 'Masculinity, Violence and War' in P. Patton and R. Poole (eds): *War/Masculinity*. Sydney: Intervention Publications.

Connell R. 1987: *Gender and Power*. Cambridge: Polity Press.

Cooley, M. 1980: *Architect or Bee? The Human/Technology Relationship*. Slough: Langley Technical Services.

Corea, G. et al 1985: *Man-Made Women: How new reproductive technologies affect women*. London: Hutchinson.

Cowan, R. S. 1976: 'The "Industrial Revolution" in the Home: Household Technology and Social Change in the Twentieth Century'. *Technology and Culture*, 17, pp. 1-23. Reprinted as 'The Industrial Revolution in the Home', in MacKenzie and Wajcman (eds) 1985. pp. 181-201.

Cowan, R. S. 1979: 'From Virginia Dare to Virginia Slims: Women and Technology in American Life'. *Technology and Culture*, 20, 1, pp. 51-63.

Cowan, R. S. 1983: *More Work for Mother: The Ironies of Household Technology from the Open Hearth to the Microwave*. New York: Basic Books.

Crompton, R. and Jones, G. 1984: *White-Collar Proletariat: Deskilling and Gender in Clerical Work*. London: Macmillan.

Crowe, C. 1987: 'Women Want It: In vitro Fertilization and Women's Motivations for Participation' in Spallone, P. and Steinberg, D. (eds): *Made To Order: The Myth of Reproductive and Genetic Progress*. Oxford: Pergamon Press.

Davidoff, L. 1976: 'The rationalization of housework' in Barker, D. and Allen, S. (eds): *Dependence and Exploitation in Work and Marriage* London: Longman.

Davidoff, L. and Hall, C. 1987: *Family Fortunes: Men and women of the English middle class, 1780–1850*. London: Hutchinson.

Deem, R. (ed.) 1980: *Schooling for Women's Work*. London: Routledge & Kegan Paul.

Dickson, D. 1984: *The New Politics of Science*. New York: Pantheon.

Dinnerstein, D. 1976: *The Mermaid and the Minotaur: Sexual Arrangements and Human Malaise*. New York: Harper & Row.

Donnison, J. 1977: *Midwives and Medical Men: A History of Inter-Professional Rivalries and Women's Rights*. London: Heinemann.

Doyal, L. 1979: *The Political Economy of Health*. London: Pluto Press.

Dugdale, A. 1988: 'Keller's Degendered Science'. *Thesis Eleven, 21*, pp. 117–28.

Easlea, B. 1981: *Science and Sexual Oppression: Patriarchy's Confrontation with Woman and Nature*. London: Weidenfeld & Nicolson.

Easlea, B. 1983: *Fathering the Unthinkable: Masculinity, Scientists and the Nuclear Arms Race*. London: Pluto Press.

Ehrenreich, B. and English, D. 1975: 'The Manufacture of Housework'. *Socialist Revolution*, 26, pp. 5–40.

Ehrenreich, B. and English, D. 1976: *Witches, Midwives and Nurses: A History of Women Healers*. London: Writers & Readers.

Ehrenreich, B. and English, D. 1979: *For Her Own Good: 150 Years of the Experts' Advice to Women*. London: Pluto Press.

Eisenstein, H. 1984: *Contemporary Feminist Thought*. London: Allen and Unwin.

Elger, T. 1987: 'Review Article: Flexible Futures? New Technology and the Contemporary Transformation of Work'. *Work, Employment and Society*, 1, 4, pp. 528–40.

Elshtain, J. 1987: *Women and War*. New York: Basic Books.

Enloe, C. 1980: *Ethnic Soldiers: State Security in Divided Societies*. Athens, Georgia: University of Georgia Press.

Enloe, C. 1983: *Does Khaki Become You?: The Militarisation of Women's Lives*. London: Pluto Press.

Epstein, C. Fuchs 1988: *Deceptive Distinctions: Sex, Gender, and the Social Order*. New Haven, Connecticut: Yale University Press.

Equal Opportunities Commission 1985: *Infotech and Gender: an Overview*.

Fagerhaugh, S., Strauss, A., Suczek, B., Wiener, C. (eds) 1987: *Hazards in Hospital Care: Ensuring Patient Safety*. San Francisco, California: Jossey-Bass.

Farrant, W. 1985: 'Who's for Amniocentesis? The Politics of Prenatal Screening' in Homans, H. (ed.). 1985.

Farrar, A. 1985: 'War: Machining Male Desire' in P. Patton and R. Poole (eds): *War/Masculinity*. Sydney: Intervention Publications.

Faulkner, W. and Arnold, E. (eds) 1985: *Smothered by Invention: Technology in Women's Lives*. London: Pluto Press.

Faulkner, W. 1985: 'Medical technology and the right to heal' in Faulkner, W. and Arnold, E. ibid.

Fava, S. 1980: 'Women's Place in the New Suburbia' in G. Wekerle, R. Peterson, and D. Morley (eds): *New Space for Women*. Boulder, Colorado: Westview Press.

Fee, E. 1981: 'Women's Nature and Scientific Objectivity' in M. Lowe and R. Hubbard (eds): *Woman's Nature: Rationalizations of Inequality*. New York: Pergamon Press.

Firestone, S. 1970: *The Dialectic of Sex*. New York: William Morrow and Co.

Fischer, C. 1988: ' "Touch Someone": The Telephone Industry Discovers Sociability'. *Technology and Culture*, 29, 1, pp. 32–61.

Florman, S. 1976: *The Existential Pleasures of Engineering*. New York: St Martin's.

Forty, A. 1986: *Objects of Desire: Design and Society 1750–1980*. London: Thames and Hudson.

Foucault, M. 1973: *The Birth of the Clinic*. London: Tavistock.

Fuchs, V. 1968: 'The Growing Demand for Medical Care'. *New England Journal of Medicine*, 279 (4), 190–95.

Game, A. and Pringle, P. 1983: *Gender At Work*. Sydney, Allen & Unwin.

Gershuny, J. 1978: *After Industrial Society: The Emerging Self-Service Economy*. London: Macmillan.

Gershuny, J. 1983: *Social Innovation and the Division of Labour*. Oxford: Oxford University Press.

Gershuny, J. 1985: 'Economic Development and Change in the Mode of Provision of Services' in Redclift, N. and Minigione, E. (eds): *Beyond Employment: Household, Gender and Subsistence*. Oxford: Basil Blackwell.

Gershuny, J. and Robinson, J. 1988: 'Historical Changes in the Household Division of Labour'. Unpublished manuscript.

Giddens, A. 1984: *The Constitution of Society*. Berkeley, California: University of California Press.

Gilligan, C. 1982: *In a Different Voice: Psychological Theory and Women's Development*. Cambridge, Mass.: Harvard University Press.

Giordano, R. 1988: 'The Social Context of Innovation: A Case History of the Development of COBOL Programming Language'. Columbia University, Department of History.

Gittins, D. 1982: *Fair Sex: Family size and structure*. London: Hutchinson.

Gordon, L. 1977: *Woman's Body, Woman's Right: A Social History of Birth Control in America*. Harmondsworth: Penguin.

Gorz, A. 1982: *A Farewell to the Working Class*. London: Pluto.

Gray, A. 1987: 'Behind closed doors: video recorders in the home' in Baehr, H. and Dyer, G. (eds): *Boxed In: Women and Television*. London: Routledge & Kegan Paul.

Greenbaum, J. 1979: *In the Name of Efficiency: Management theory and shopfloor practice in data processing work*. Philadelphia, Pennsylvania: Temple University Press.

Greenwood, K. and King, L. 1981: 'Contraception and Abortion' in Cambridge Women's Studies Group (ed): *Women in Society*. London: Virago.

Griffin, S. 1983: in L. Caldecott and S. Leland (eds): *Reclaim the Earth*. London: The Women's Press.

Griffiths, D. 1985: 'The exclusion of women from technology' in Faulkner and Arnold, 1985.

Hacker, S. 1981: 'The Culture of Engineering: Woman, Workplace and Machine'. *Women's Studies International Quarterly*, 4, pp. 341–53.

Hacker, S. 1989: *Pleasure, Power and Technology*. Boston, Massachussetts: Unwin Hyman.

Haddon, L. 1988: 'The Roots and Early History of the British Home Computer Market: Origins of the Masculine Micro'. PhD thesis, Management School, Imperial College. University of London.

Hanmer, J. 1985: 'Transforming Consciousness: Women and the new reproductive technologies' in Corea, G. et al 1985.

Haraway, D. 1985: 'A Manifesto for Cyborgs: Science, Technology and Socialist Feminism in the 1980s'. *Socialist Review*, 80, 15, pp. 65–107.

Harding, S. 1986: *The Science Question in Feminism*. New York: Cornell University Press.

Hardyment, C. 1988: *From Mangle to Microwave: The Mechanization of the Household*. Cambridge, Polity Press.

Harman, E. 1983: 'Capitalism patriarchy and the city' in C. Baldock and B. Cass (eds) *Women, Social Welfare and the State in Australia*. Sydney: Allen and Unwin.

Hartmann, H., Kraut, R., and Tilly, L. (eds) 1986, 1987: *Computer Chips and Paper Clips: Technology and Women's Employment*, Volumes 1 and 2. Washington DC: National Academy Press.

Hartsock, N. 1983: 'The Feminist Standpoint: Developing the Ground for a Specifically Feminist Historical Materialism' in S. Harding and M. Hintikka (eds) *Discovering Reality: Feminist Perspectives on Epistemology, Metaphysics, Methodology and Philosophy of Science*. Dordrecht: Reidel.

Harvey, D. 1989: *The Condition of Postmodernity*. Oxford: Basil Blackwell.

Hayden, D. 1980: 'Redesigning the Domestic Workplace' in G. Wekerle, R. Peterson and D. Morley (eds): *New Space for Women*. Boulder, Colorado: Westview Press.

Hayden, D. 1982: *The Grand Domestic Revolution: A History of Feminist Designs for American Homes, Neighborhoods, and Cities*. Cambridge, Massachussetts: MIT Press.

Himes, N. 1936: *Medical History of Contraception*. Baltimore, Maryland: Williams & Wilkins.

Homans, H. (ed.) 1985: *The Sexual Politics of Reproduction*. London: Gower.

Hoyles, C. (ed.) 1988: *Girls and Computers*, Bedford Way Papers 34. London: Institute of Education.

Hughes, M., Brackenridge, A., Bibby, A., Greenhaugh, P. 1988: 'Girls, Boys and Turtles: gender effects in young children learning with Logo' in C. Hoyles 1988.

Huxley, M. 1988: 'Feminist Urban Theory: Gender, Class and the Built Environment'. *Transition*, 25, pp. 39–43.

Jacobs, J. 1962: *The Death and Life of Great American Cities*. London: Jonathan Cape.

Jewson, N. 1976: 'The disappearance of the sick man from medical cosmology, 1770–1870'. *Sociology*, vol. 10, no. 2, pp. 225–44.

Jones, B. 1982: *Sleepers Wake!* Melbourne: Oxford University Press.

Jordanova, L. J. 1980: 'Natural Facts: A Historical Perspective on Science and Sexuality' in C. MacCormack and M. Strathern (eds) *Nature, Culture and Gender*. Cambridge: Cambridge University Press.

Jordanova, L. 1987: 'Gender, Science and Creativity' in M. McNeil (ed.) *Gender and Expertise*. London: Free Association Books.

Jordanova, L. 1989: *Sexual Visions: Images of Gender in Science and Medicine between the Eighteenth and Twentieth Centuries*. London, Harvester Wheatsheaf.

Keller, E. Fox 1983: *A Feeling for the Organism: The Life and Work of Barbara McClintock*. San Francisco, California: Freeman.

Keller, E. Fox 1985: *Reflections on Gender and Science*. New Haven, Connecticut: Yale University Press.

Kelly, A. (ed.) 1981: *The Missing Half: Girls and Science Education*. Manchester: Manchester University Press.

Kennedy, M. 1981: 'Toward a Rediscovery of "Feminine" Principles In Architecture and Planning'. *Women's Studies International Quarterly*, 4, 1, pp. 75–81.

Kidder, T. 1982: *The Soul of a New Machine*. Harmondsworth: Penguin.

Klein, R. 1985: 'What's "New" about the "New" reproductive technologies?', in Corea et al 1985.

Knorr-Cetina, K. and Mulkay, M. (eds) 1983: *Science Observed: Perspectives in the Social Studies of Science*. London: Sage.

Kolata, G. 1984: 'Equal Time for Women'. *Discover*, January, 1984.

Kraft, P. 1977: *Programmers and Managers: The Routinization of Computer Programming in the United States*. New York: Springer Verlag.

Kraft, P. 1979: 'The routinization of computer programming'. *Sociology of Work and Occupations*, 6, pp. 139–55.

Kraft, P and Dubnoff, S. 1986: 'Job Content, Fragmentation and Control in Computer Software Work'. *Industrial Relations*, 25, 2, pp. 184–96.

Kramarae, C. (ed.) 1988: *Technology and Women's Voices* New York: Routledge & Kegan Paul.

Kranakis, E. 1989: 'Social Determinants of Engineering Practice: A Comparative View of France and America in the Nineteenth Century'. *Social Studies of Science*, Vol. 19, pp. 5–70.

Kuhn, T. 1970: *The Structure of Scientific Revolutions*. Chicago, Illinois: Chicago University Press.

Kuhrig, H. 1978: *Zur gesellshaftlischen Stellung der Frau in der DDR*. Leipzig: Verlag fuer die Frau.

Law, J. 1987: 'Review Article: The structure of sociotechnical engineering – a review of the new sociology of technology'. *The Sociological Review*, 35, pp. 404–25.

Lazonick, W. 1979: 'Industrial Relations and Technical Change: The Case of the Self-Acting Mule'. *Cambridge Journal of Economics*, 3 pp. 231–62.

Leavitt, J. W. 1986: *Brought to Bed: Childbearing in America 1750 to 1950*. New York: Oxford University Press.

Lewis, J. 1984: *Women in England 1870–1950: Sexual Divisions and Social Change*. Brighton: Wheatsheaf.

Lewis, L. 1987: 'Females and Computers: Fostering Involvement' in B. Drygulski Wright et al. (eds) *Women, Work, and Technology*. Ann Arbor, Michigan: University of Michigan Press.

Liff, S. 1986: 'Technical Change and Occupational Sex-typing' in D. Knights and H. Willmott (eds): *Gender and the Labour Process*. Aldershot: Gower.

Liff, S. 1988: 'Gender, Office Work and Technological Change'. Working Paper 176, Management Studies, Loughborough University of Technology.

Linn, P. 1985: 'Microcomputers in Education: Dead and Living Labour' in T. Solomonides and L. Levidow (eds): *Compulsive Technology: Computers as Culture*. London: Free Association Books.

Linn, P. 1987: 'Socially Useful Production'. *Science as Culture*, 1, pp. 105–138.

Lloyd, G. 1986: 'Selfhood, war and masculinity' in C. Pateman and E. Gross (eds): *Feminist Challenges*. Sydney: Allen and Unwin.

Lloyd, G. 1984: *The Man of Reason*. London: Methuen.

Lovegrove, G. and Hall, W. 1987: 'Where have all the girls gone?'. *University Computing*, 9, 207–10.

Macintyre, S 1977: 'Childbirth: the myth of the Golden Age'. *World Medicine*, 15 June, 17–22.

MacKenzie, D. 1984: 'Marx and the Machine', *Technology and Culture*, 25, pp. 473–502.

MacKenzie, D. and Wajcman, J. (eds) 1985: *The Social Shaping of Technology*. Milton Keynes: Open University Press.

MacLeod, C. 1987: 'Accident or Design? George Ravenscroft's Patent and the Invention of Lead-Crystal Glass'. *Technology and Culture*, 28, 4, pp. 776–803.

McFarlane, B. 1984: 'Homes fit for heroines: housing in the twenties' in Matrix: *Making Space: Women and the Man-Made Environment*. London: Pluto Press.

McGaw, J. 1982: 'Women and the History of American Technology'. *Signs:*

Journal of Women in Culture and Society, 7, pp. 798–828.
McKinlay, J. B. 1981: 'From "Promising Report" to "Standard Procedure" – Seven Stages in the Career of a Medical Innovation'. *Milbank Memorial Fund Quarterly*, 59, 3, 374–411.
McLaren, A. 1984: *Reproductive Rituals: The perception of fertility in England from the sixteenth century to the nineteenth century*. London: Methuen.
McNeil, M. (ed.) 1987: *Gender and Expertise*. London: Free Association Books.
McNeil, M., Varcoe, I. and Yearley, S. (eds) 1990: *The New Reproductive Technologies*. London: MacMillan.
Mannheim, K. 1953: *Essays on Sociology and Social Psychology*. London: Routledge and Kegan Paul.
Marcuse, H. 1968: *Negations*. London: Allen Lane.
Margolis, M. 1985: *Mothers and Such: Views of American Women and Why They Changed*. Berkeley, California: University of California Press.
Martin, E. 1987: *The Woman in the Body: A Cultural Analysis of Reproduction*. Boston, Massachussetts: Beacon Press.
Marx, K. 1975: *Early Writings*. New York: Vintage Books.
Matrix (eds) 1984: *Making Space: Women and the Man-Made Environment*. London: Pluto Press.
Merchant, C. 1980: *The Death of Nature: Women, Ecology and the Scientific Revolution*. New York: Harper and Row.
Mies, M. 1987: 'Why do we Need all this? A Call Against Genetic Engineering and Reproductive Technology' in Spallone, P. and Steinberg, D. (eds): *Made To Order: The Myth of Reproductive and Genetic Progress*. Oxford: Pergamon Press.
Miles, I. 1988: *Home Informatics: Information Technology and the Transformation of Everyday Life*. London: Pinter Publishers.
Morley, D. 1986: *Family Television: Cultural Power and Domestic Leisure*. London, Comedia.
Moyal, A. 1989: 'The Feminine Culture of the Telephone'. *Prometheus*, 7, 1, pp. 5–31.
Mumford, L. 1961: *The City in History*. New York: Harcourt, Brace and World.
Newman, E. 1985: 'Who controls birth control?' in Faulkner, W. and Arnold, E. (eds) 1985.
Newman, P. 1988: 'Australian Cities at the Crossroads'. *Current Affairs Bulletin*, 65, 7, pp. 4–15.
Noble, D. 1984: *Forces of Production: A Social History of Industrial Automation*. New York: Knopf.
Oakley, A. 1974: *The Sociology of Housework*. London: Martin Robertson.
Oakley, A. 1976: 'Wisewoman and Medicine Man: Changes in the Management of Childbirth' in Mitchell, J. and Oakley, A. (eds): *The Rights and Wrongs of Women*. Harmondsworth: Penguin.

Oakley, A. 1987: 'From Walking Wombs to Test-Tube Babies' in M. Stanworth (ed.) 1987.

O'Donnell, C. 1984: *The Basis of the Bargain: Gender, Schooling and Jobs*. Sydney: Allen and Unwin.

Pacey, A. 1983: *The Culture of Technology*. Oxford: Basil Blackwell.

Palm, R. and Pred, A. 1978: 'The Status of American Women: A Time-Geographic View' in D. Lanegran and R. Palm (eds): *An Invitation to Geography*. New York: McGraw-Hill.

Parsons, T. 1956: 'The American family: its relations to personality and the social structure' in T. Parsons and R. Bales: *Family, Socialisation and Interaction Process*. London, Routledge & Kegan Paul.

Petchesky, R. 1987: 'Foetal Images: the Power of Visual Culture in the Politics of Reproduction' in M. Stanworth (ed.) 1987.

Pfeffer, N. 1985: 'The Hidden Pathology of the Male Reproductive System' in Homans, H. (ed.) 1985.

Pfeffer, N. 1987: 'Artificial Insemination, In-vitro Fertilization and the Stigma of Infertility' in M. Stanworth (ed.) 1987.

Pickup, L. 1988: 'Hard to Get Around: A Study of Women's Travel Mobility' in J. Little, L. Peake, and P. Richardson (eds): *Women in Cities: Gender and the Urban Environment*. London: Macmillan.

Pollert, A. 1988: 'Dismantling flexibility'. *Capital and Class*, 34, pp. 42–75.

Pollack, S. 1985: 'Sex and the Contraceptive Act' in Homans, H. (ed.) 1985.

The Prince of Wales 1989: *A Vision of Britain: A Personal View of Architecture*. London: Doubleday.

Purcell, K. 1988: 'Gender and the Experience of Employment' in Gallie, D. (ed.): *Employment in Britain*. Oxford: Basil Blackwell.

Pursell, C. 1981: 'Women Inventors in America'. *Technology and Culture*, 22, 3, pp. 545–9.

Rakow, L. 1988: 'Women and the telephone: the gendering of a communications technology' in Kramarae, C. (ed.) 1988.

Rakusen, J. and Davidson, N. 1982: *Out of Our Hands: What Technology does to Pregnancy*. London: Pan.

Ravetz, A. 1965: 'Modern Technology and an Ancient Occupation: Housework in Present-Day Society'. *Technology and Culture*, 6, pp. 256–60.

Ravetz, A. 1987: 'Housework and Domestic Technologies' in M. McNeil (ed.) 1987.

Ravetz, A. 1989: 'A View from the Interior' in Attfield, A. & Kirkham, P. (eds): *A View from the Interior: Feminism, Women and Design*. London: The Women's Press.

Reiger, K. 1985: *The Disenchantment of the Home: Modernizing the Australian Family 1880–1940*. Melbourne: Oxford University Press.

Reiger, K. 1986: 'At Home with Technology'. *Arena*, 75, pp. 109–23.

Reiser, S. 1978: *Medicine and the Reign of Technology*. Cambridge: Cambridge University Press.

Renner, M. 1988: *Rethinking the Role of the Automobile*. Worldwatch Paper 84, Washington, June 1988.

Richards, E. and Schuster, J. 1989: 'The Feminine Method as Myth and Accounting Resource: A Challenge to Gender Studies and Social Studies of Science'. *Social Studies of Science*, 19, pp. 697–720.

Richards, M. P. M. (ed.) 1978: *The Hazards of the New Obstetrics*. London: Pitman Medical.

Richards, M. and Light, P. (eds) 1986: *Children of Social Worlds*. Cambridge: Polity Press.

Rose, H. 1983: 'Hand, Brain, and Heart: A Feminist Epistemology for the Natural Sciences'. *Signs*, 9, 1 (Fall), pp. 73–90.

Rosenberg, N. 1979: 'Technological Interdependence in the American Economy'. *Technology and Culture*, vol. 20, no. 1 pp. 25–50.

Rothschild, J. (ed..) 1983: *Machina Ex Dea: Feminist Perspectives on Technology*. New York: Pergamon Press.

Rowland, R. 1985: 'Motherhood, patriarchal power, alienation and the issue of "choice" ' in Corea et al. (eds) 1985.

Rowland, R. 1988: *Woman Herself – A Transdisciplinary Perspective on Women's Identity*. Melbourne: Oxford University Press.

Ruddick, S. 1983: 'Pacifying the Forces: Drafting Women in the Interests of Peace'. *Signs*, 8, 3, pp. 471–89.

Saegert, S. 1980: 'Masculine Cities and Feminine Suburbs: Polarized Ideas, Contradictory Realities'. *Signs*, 5, 3, pp. 96–111.

Sayre, A. 1975: *Rosalind Franklin and DNA: A Vivid View of What It Is Like to Be a Gifted Woman in an Especially Male Profession*. New York: W. W. Norton & Co.

Scharff, V. 1988: 'Putting wheels on women's sphere' in Kramarae, C (ed.) 1988.

Schiebinger, L. 1987: 'The History and Philosophy of Women in Science: A Review Essay'. *Signs*, 12, 2, pp. 305–32.

Schuster, J. S. and Yeo, R. 1986: *The Politics and Rhetoric of Scientific Method*. Dordrecht: Reidel.

Segal, L. 1987: *Is the Future Female? Troubled Thoughts on Contemporary Feminism*. London: Virago.

Shorter, E. 1983: *A History of Women's Bodies*. London: Allen Lane.

Sirageldin, I. 1969: *Non-Market Components of National Income*. Ann Arbor, Michigan: University of Michigan Survey Research Center.

Soja, E. 1989: *Postmodern Geographies*. London: Verso.

Spallone, P. 1987: 'Reproductive Technology and the State: The Warnock Report and its Clones' in Spallone, P. and Steinberg, D. (eds) *Made to Order: The Myth of Reproductive and Genetic Progess*. Oxford: Pergamon.

Stanley, A. forthcoming: *Mothers of Invention*. Metuchen, New Jersey: Scarecrow Press.

Stanworth, M. (ed.) 1987: *Reproductive Technologies: Gender, Motherhood and Medicine*. Cambridge: Polity Press.

Staudenmaier, J. 1985: *Technology's Storytellers*. Cambridge, Mass.: MIT Press.

Stein, D. 1985: *Ada: A Life and Legacy*. Cambridge, Mass.: MIT Press.

Strasser, S. 1982: *Never Done: A History of American Housework*. New York: Pantheon.

Strasser, S. 1980: 'An Enlarged Human Existence? Technology and Household Work in Nineteenth-Century America' in Berk S. F. (ed.): *Women and Household Labor*. Beverly Hills, California: Sage.

Strathern, M. 1980: 'No Nature, No Culture: the Hagen Case' in C. MacCormack and M. Strathern (eds): *Nature, Culture and Gender*. Cambridge: Cambridge University Press.

Sutherland, R. and Hoyles, C. 1988: 'Gender Perspectives on Logo Programming in the Mathematics Curriculum' in C. Hoyles (ed.) 1988.

Thompson, P. 1983: *The Nature of Work: An Introduction to Debates on the Labour Process*. London: Macmillan.

Thrall, C. 1982: 'The Conservative Use of Modern Household Technology'. *Technology and Culture*, 23, pp. 175–94.

Torre, S. (ed.) 1977: *Women in American Architecture*. New York: Whitney Library of Design.

Toffler, A. 1980: *The Third Wave*. London: Collins/Pan.

Trescott, M. M. (ed.) 1979: *Dynamos and Virgins Revisited: Women and Technological Change in History*. Metuchen, New Jersey: Scarecrow Press.

Turkle, S. 1984: *The Second Self: Computers and the Human Spirit*. London: Granada.

Turkle, S. and Papert, S. forthcoming: 'Epistemological Pluralism: Styles and Voices within the Computer Culture'. *Signs*.

Vanek, J. 1974: 'Time Spent on Housework'. *Scientific American*, 231, 5, pp. 116–20.

Wajcman, J. 1983: *Women in Control: Dilemmas of a Workers Cooperative*. Milton Keynes: Open University Press.

Wajcman, J. and Probert, B. 1988: 'New Technology Outwork' in Willis, E. (ed.): *Technology and the Labour Process: Australian Case Studies*. Sydney: Allen and Unwin.

Walkerdine, V. and The Girls and Mathematics Unit 1989: *Counting Girls Out*. London: Virago.

Webster, J. 1989: *Office Automation: The Labour Process and Women's Work in Britain*. Hemel Hempstead: Wheatsheaf.

Weizenbaum, J. 1976: *Computer Power and Human Reason: From Judgment to Calculation*. Harmondsworth: Penguin.

Wilkinson, B. 1983: *The Shopfloor Politics of New Technology*. London: Heinemann.

Willis, P. 1977: *Learning to Labour: How Working Class Kids Get Working Class Jobs*. London: Saxon House.

Winner, L. 1980: 'Do Artifacts Have Politics?' *Daedalus*, 109, pp. 121–36.

Wolfe, T. 1980: *The Right Stuff*. London: Jonathan Cape.

Women and Transport Forum 1988: 'Women on the move: how public is public transport?' in Kramarae, C. (ed.) 1988.

Wright, G. 1981: *Building the Dream: A Social History of Housing in America*. New York: Pantheon.

Wyatt, S., Thomas, G. and Miles, I. 1985: 'Preliminary Analysis of the ESRC 1983/4 Time Budget Data'. Science Policy Research Unit, University of Sussex.

Yoxen, E. 1985: 'Licensing Reproductive Technologies?' in Issues in Radical Science. *Radical Science Journal*, 17, 138–48.

Yoxen, E. 1986: *The Gene Business: Who Should Control Biotechnology?* London: Free Association Books.

Yoxen, E. 1987: 'Seeing with Sound: A Study of the Development of Medical Images' in W. Bijker, et al. (eds) 1987.

Zimmerman, J. (ed.) 1983: *The Technological Woman: Interfacing with Tomorrow*. New York: Praeger.

Zimmerman, J. 1986: *Once Upon the Future: A Woman's Guide to Tomorrow's Technology*. New York: Pandora.

Index

architecture
  feminist perspectives 120–6
  and modernism 113–14, 121
  see also housing; urban planning
automation
  of machine tools 46–8
  office work 29–33
  typesetting 49–51
automobiles
  alternatives to 128
  culture of 133–5
  effects on urban space 126–7
  gender differences in use 129–31
  manufacture 126
  see also transportation system

Baran, Barbara 32
biotechnology 73
birth control
  see contraceptive technology
Braverman, Harry 20
  see also labour process debate
buying patterns 84

car
  culture of 133–5
  see also automobiles
childbirth management
  male domination of 63–73
  midwives 63–5
class
  divisions 43–6
  intersections with gender and race 11, 24, 36, 52, 61, 120, 143
  shape technology 43–8
clerical work
  decentralization 40
  deskilling of 29–32
  feminization 32
Cockburn, Cynthia 19–21, 36, 38–9, 49–51, 89, 146
cognitive sex difference 7, 155–8
Cohn, Carol 140
computers
  as games 153–5
  male culture of 141–2, 144–6, 153–5
  programming styles 155–8
  at school 150–3
Connell, Bob 143
contraceptive technology 74–8
Cooley, Mike 18
Cowan, Ruth Schwartz 83–7, 91, 92, 97–9, 101–2
craft unionism
  and women 21, 32–3, 36, 38–9, 49–52
  see also trade unions
craft work
  and technology 21, 36, 39, 43–52
culture of work
  of engineering 34, 48, 102, 145–6
  of masculinity 38–40, 138–49

Davidoff, Leonore and Hall, Catherine 112–13
design choice
  and profit 100–5
  sex bias 49–52
deskilling
  craft workers 43–8
  office workers 29–33
domestic science movement 85, 98, 113–14
  see also housework
domestic servants 85, 104, 116
domestic technology
  design of 99–105
  and economics 91–5, 100–4
  electrical appliances 103–4
  gender specialization of 87–90
  washing machines 92–3
  see also housework; refrigerators

Easlea, Brian 4, 138–9
economics
  and domestic technology 91–5, 100–1, 102–4
  shape technology 43, 48, 72–3, 128
electronic homework 40–2
engineers
  culture of 34, 102, 145–6
Enloe, Cynthia 147–9

family
  consumption 84
  and industrialization 83
  and work 33
  see also housework; housing; single-family household
femininity 2–3, 9, 19, 38, 152
Feminist International Network of Resistance to Reproductive and Genetic Engineering (FINRRAGE) 58–63
feminist theory
  cultural feminism 9
  eco-feminism 6–7, 18, 135, 163
  liberal feminism 8
  postmodernism 11
  radical feminism 9–10, 58–60
  socialist feminism 20
Firestone, Shulamith 13, 56, 58
Fischer, Claude 104–5
flexible specialization 28
forceps 64–5
Forty, Adrian 104
fragmentation of work
  see deskilling

Gabe, Frances 102
gendering of jobs 33–4, 37–40
Gershuny, Jonathan 91–4
Gilligan, Carol 148, 157
Gilman, Charlotte Perkins 125
Gordon, Linda 74–5
Gray, Ann 90

Hacker, Sally 145
hackers 141–2, 144
Haddon, Leslie 155
Harding, Sandra 2, 4, 10–11
Hayden, Dolores 124–5
homeworking
  see electronic homework
housework
  domestic science movement 85, 98
  gender specialization of 87–90
  industrial revolution of 83–7
  socialization of 96–8, 125
  time spent on 81–2, 83–5, 91–4
  see also domestic technology
housing
  public housing 114–15
  suburban housing 118–20
  in the twentieth century 113–20
  in the Victorian period 112–13

industrial conflict
  machinists 46–8
  spinners 44–5
  and technological change 43–8
  typesetters 49–51
  see also trade unions

infertility treatment
medical interests 68
insurance industry
automation of 32
international division of labour 13, 40
invention
and social process 22-4
and women 15-17, 22
in-vitro fertilization (IVF) 57-8, 61-2

Jordanova, Ludi 6, 67

Keller, Evelyn Fox 2, 7
Kennedy, Margrit 121-4
Kidder, Tracy 141
kitchen
design of 116-17
location of 111, 113-14
kitchenless houses 96, 125
know-how and skill 38-9

labour process debate 20, 30, 43
Lazonick, William 45
Leavitt, Judith Walzer 66
leisure technology
television 90, 94
video 90
see also telephone
Lloyd, Genevieve 148

machine tool automation
managerial control 46-8
machinery
and control of labour 43-8
see also automation
Martin, Emily 67-8
Marx, Karl 1, 20
masculinity
and engineering 102, 145-6
forms of 39-40, 143-6
and technical culture 21, 38-40, 141-6
and war 18, 138-41, 146-9
matrix 120, 124
McNeil, Maureen 5, 11
medical knowledge
and gender symbolism 67-8
medical profession
male dominance of 63-6
role of technology 64-6, 69-72
medical technology
see reproductive technologies
midwives 63-5
Mies, Maria 59
Miles, Ian 95
military 146-9
military technology 149-64
see also warfare
modernist architecture 113-14, 121
Morley, David 90
Moses, Robert 133
Mumford, Lewis 18, 123

Noble, David 22, 46-8

Oakley, Ann 66, 69, 71, 81
office automation 29-33
see also clerical work

Pacey, Arnold 18
Peirce, Melusina Fay 125
post-industrial theory 27-8, 82
printing industry 35-6
proletarianization
see deskilling

race
intersections with class and gender 11, 24, 36, 52, 61, 120, 143
radical science movement 3, 20
refrigerators
design choice 101-2
Reiger, Kereen 98-9
reproductive technologies
birth control 74-8
forceps 64-5
in-vitro fertilization 57-8, 61-2
Rose, Hilary 8

science
barriers to women's participation 2–3
feminist critique of 1–12
and gender symbolism 5–6
and women's values 8–12
science–technology relationship 13–14
Segal, Lynne 9, 148, 157
self-service economy 91–5
sex segregation 29–34
sex typing of jobs 33–4
see also gendering of jobs
single-family household
and domestic technology 95–9
and housing, architecture 116–20
skill
gender bias in 37–8
see also craft work; deskilling
spinning technology
and labour control 44–5
and skilled workers 44–6
Stanworth, Michelle 60–1
stethoscope
invention of 70

technological determinism 20–4, 46, 54–5, 74, 82, 111, 162
technology
culture of 21, 38–40, 137–49
definition of 14–17, 137, 162
history of 15–17, 23
and science 13–14
sociology of 23–4, 162–3
and women's values 17–19
telephone
gendered use 105
history of 104–5
telework
see electronic homework
Third World ix, 13, 40, 61, 76
trade unions
exclusion of women 32, 49–52
and skill 37–40
and technology 43–52
transportation system 84, 126–8
see also automobiles
travel patterns
sex differences 129–31
Turkle, Sherry 141–2, 144, 156–7
twilight-sleep movement 65–6
typesetting technology 49–51

ultrasound imaging 71
urban planning 119
see also architecture

Victorian values 112–13

warfare
ideology of 138–41
language of 140
see also masculinity
Webster, Juliet 31
Winner, Langdon 63, 133
Wolfe, Tom 142
women's movement viii, 1, 5, 8, 13, 29, 54, 56, 165
working class
divisions in 43–6
see also class

zoning 119–20